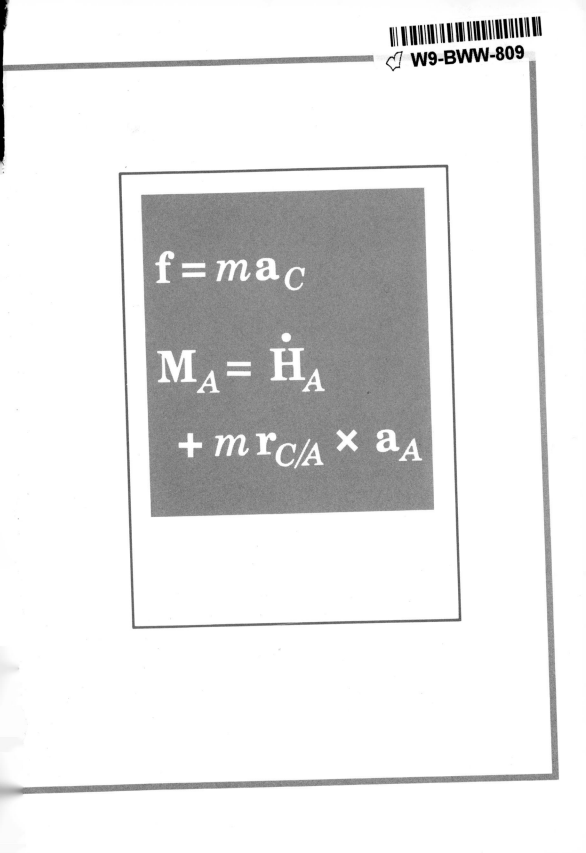

$$f = m\mathbf{a}_C$$

$$M_A = \dot{H}_A$$

$$+ m\mathbf{r}_{C/A} \times \mathbf{a}_A$$

APPLIED MECHANICS:
DYNAMICS

APPLIED MECHANICS:
DYNAMICS

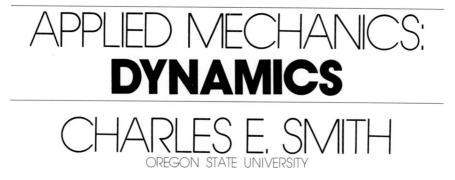

CHARLES E. SMITH
OREGON STATE UNIVERSITY

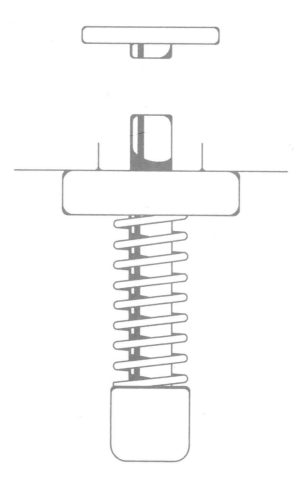

JOHN WILEY & SONS, INC.
New York   London   Sydney   Toronto

This book was set in Baskerville by Progressive Typographers.
It was printed and bound by Halliday Lithograph. The drawings
were designed and executed by John Balbalis with the assistance
of the Wiley Illustration Department. The text and cover were
designed by Suzanne G. Bennett. The copyeditor was Vivian
Kahane. The production supervisor was Kenneth Ekkens.

Library of Congress Cataloging in Publication Data:

Smith, Charles Edward, 1932–
  Applied mechanics.

  Includes bibliographies and indexes.
  CONTENTS: v. 1. Statics   v. 2. Dynamics
v. 3. More dynamics.
  1. Mechanics, Applied.   I. Title.
TA350.S57   620.1   75-44021
ISBN 0-471-80178-X (v. 2)

Printed in the United States of America

10 9 8 7 6 5 4 3 2 1

PREFACE

This volume presents analyses of kinematics and kinetics of particles, systems of particles, and rigid bodies, with which every engineer must be familiar. All of the material is essential, with the exception of Sections 11-4 and 13-5, and adequate coverage requires a minimum of three quarter hours. If numerical integration is covered and thorough attention is given to the essential topics, a course of four semester hours can be given.

A proficiency at resolving force systems and drawing free-body diagrams, and a working understanding of differentiation, integration, and vector algebra are the expected background for studying this material.

The emphasis is on the application of fundamentals, and problem solutions that have been developed in the past are used primarily for illustration. For example, instead of a chapter on central force motion per se, examples and exercises on orbital mechanics are given to develop a facility in the application of ideas of angular momentum and mechanical energy.

The practice of trying to understand the ideas of mechanics mainly in visual terms is even more crucial in dynamics than in statics. Therefore, I continue to emphasize the geometric interpretation of vectors. The symbol

$\mathbf{A} \times \mathbf{B}$ should bring forth the right hand with the fingers curled and thumb extended, but it should suggest the formula (3-22) only when a computation is required.

Because dynamics is more difficult than statics, let me repeat the comments made in the preface to *Statics,* concerning procedures for learning mechanics. Very few people can learn mechanics solely by observing the analysis carried out by someone else. At some point (the sooner the better), the student must discard the observer role and, making and correcting the inevitable mistakes, he must attempt to solve problems and carry forth derivations with increasing independence. Consequently, this book will be relatively ineffective in the lap of one sitting in an easy chair. The book must be studied at a desk with an ample supply of scratch paper at hand. The paper will be needed for attempting solutions of example problems before the given solutions are read, for carrying out steps of analysis that are omitted from the books, and for making supplemental sketches. A student who works in this way, always attempting to relate solution procedures to basic ideas, will achieve an understanding of dynamics that is satisfying as well as professionally useful. I wish him well.

Charles E. Smith

Corvallis, Oregon
July 1975

ACKNOWLEDGMENTS

There were many contributors to this project, and I thank all of them. Valuable suggestions came from many students, especially Brad Whiting and John Gale. Dr. Hans J. Dahlke pointed out a number of errors and shortcomings in *Statics*. Dr. William E. Holley, Dr. Robert W. Thresher, and Dr. Robert E. Wilson contributed significantly to *Dynamics*. I am grateful to the staff of Wiley for the editing and production. The illustrations were prepared by John Balbalis. The most eminent contributor, and a source of inspiration, is Emeritus Professor Kenneth E. Bisshopp of Rensselaer Polytechnic Institute. Finally, for nearly all of the time made available, I thank Marian, Brian, and Susan.

C.E.S.

CONTENTS

ix

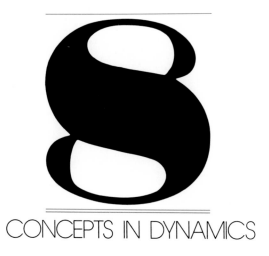

CONCEPTS IN DYNAMICS

Dynamics vis-à-vis Statics. Chapters 2 to 7 dealt with the special case of Newton's laws of motion, in which the bodies under consideration are motionless. This led us to make a fairly thorough analysis of the left-hand side of the equation $\mathbf{f} = m\mathbf{a}$ (and its rotational counterpart, which involves moments of forces). The remaining concern will be with the analysis of systems in which there are points undergoing accelerations; thus a logical next step is the achievement of a working understanding of the right-hand side of this equation and its rotational counterpart. This brings us to *kinematics*, which is the study of various aspects of motion without reference to the forces that cause it.

Kinematics is challenging and fascinating; in addition to providing a basis for design of many useful devices, it is rich with surprising features. Consider, for example, airplane A in Figure 8-1. In its steady banking turn of radius 3.7 km, its acceleration (time rate of change of velocity) turns out to be 17 m/s² to the left, even though the magnitude of its velocity is constant. In addition, to an observer in aircraft A, the other aircraft appears to have a *rearward* velocity of 108 km/h! Detailed analysis of kinematics is facilitated considerably by means of calculus of vector func-

1

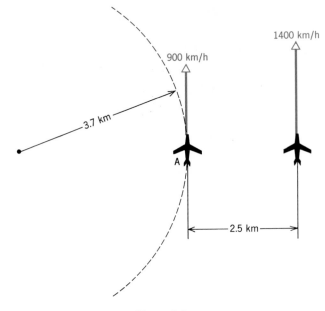

Figure 8-1

tions of one scalar variable (time), the subject of Chapter 9 and part of Chapter 14.

Just as a sound understanding of force systems enables us to write $\mathbf{f} = \mathbf{0}$ properly for motionless systems, the additional understanding of kinematics enables us to apply $\mathbf{f} = m\mathbf{a}$ properly to moving systems. However, the subsequent task of *solving* the resulting equations often presents challenges beyond those encountered in statics of rigid body systems.

The dynamics problems in which the motions are known are similar to statics problems, in that applications of Newton's laws lead to readily solvable equations for the unknown forces. Generally more difficult are dynamics problems in which some aspects of the motion are unknown. In these, the application of Newton's second law normally results in a set of differential equations, integration of which may be relatively easy or very difficult. For example, if we know that the shaft in a stationary reciprocating engine is turning at constant speed, the accelerations of the internal moving parts can be determined through kinematics, and the forces then readily evaluated. But, if the load being driven by the engine is such that the shaft speed may be fluctuating, the prediction of this irregular motion, in terms of the forces on the pistons and the characteristics of the load, may be quite difficult.

Drawing a correct free-body diagram has been found to be indispensable in the analysis of mechanical systems in static equilibrium. This practice is no less important for the analysis of systems on which the force and moment resultants are no longer zero. Nor is the procedure different in any way: the system must be clearly identified and isolated from other bodies, and all forces of interaction with bodies *outside* the system must be included in the force side of the equation.

Inertial Reference Frames.   The *law of inertia,* deduced by Galileo and incorporated by Newton as the first of his three basic postulates of mechanics, states that a material body not subjected to action by forces either remains at rest or moves in a straight line with constant speed. But implicit in the notions of "straight line" and "constant speed" is some reference frame in which these are to be reckoned. An object that appears to a passenger on a merry-go-round to be motionless will not appear motionless to a bystander on the ground. Thus Newtonian mechanics rests on the notion of a so-called *primary* reference frame, attached to "fixed stars." Any reference frame that translates at constant velocity in this primary reference frame is called an *inertial* reference frame. In the study of kinematics, we will establish that a particle that appears from the primary reference frame to be moving in a straight line at constant speed will appear from any inertial reference frame to be moving in a straight line at constant speed. In terms of observable phenomena, an inertial reference frame is one from which a completely isolated body appears to move at constant velocity.

In many situations the earth may well approximate an inertial reference frame, because the accelerations associated with its rotational and translational motion are negligible. In other situations these effects may be of sufficient magnitude to warrant their account.

Problems

8-1   A 180-lb man in a test sled is being accelerated horizontally at a rate of six times the free fall acceleration near the earth's surface. Draw a free-body diagram of the man, and evaluate (a) the resultant force on the man, and (b) the resultant reaction from the seat.

8-2   Immediately after release from a 60-lb draw, a longbow accelerates a 55-g arrow from rest. Draw free-body diagrams of (a) the bow, not including the string, (b) the bow with string, (c) the arrow, and (d) the system consisting of bow, arrow, and string, all for the instant immediately after release. Neglecting all mass except that of the arrow, estimate the acceleration of the arrow at this instant.

8-3   A freight train is accelerating up the grade at 0.328 ft/s². The mass of the engine is 145 Mg, and that of each car 35 Mg. Draw free-body diagrams of (a) the train, (b) the engine, and (c) each car. Estimate the tension in each coupler.

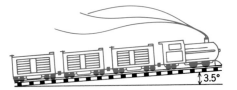

8-4   A 3700-lb automobile is accelerating along a horizontal road at 4 m/s². Draw free-body diagrams of (a) the entire car, (b) a rear wheel, and (c) the car minus its rear wheels.

8-5   Draw free-body diagrams of each block.

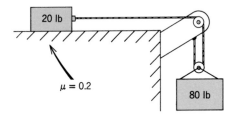

8-6   An astronaut is "floating" inside a space ship as it orbits the earth at an altitude of 180 mi. Draw a free-body diagram of the astronaut. Assuming we have not discovered a violation of Newton's laws of motion and gravitation, what is the astronaut's acceleration?

8-7   The demolition ball is moving in a circular path in a horizontal plane. Neglecting the mass of the cable, draw free-body diagrams of (a) the cable, and (b) the ball.

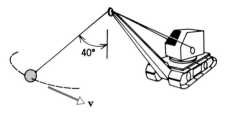

8-8   For the state immediately after the workman *B* releases his end of the plank, draw a free-body diagram of the plank.

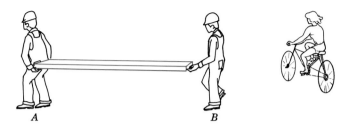

A
B

8-9 The right-hand aircraft in Figure 8-1 travels in a straight line at constant speed as observed from the earth. Sketch the path of this aircraft as it appears from a reference frame attached to aircraft A.

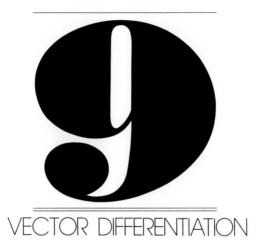

VECTOR DIFFERENTIATION

In this chapter we consider vectors that are continuous, differentiable functions of a single scalar variable t. This variable will be interpreted as *time* in most of our applications in dynamics.

9–1

DEFINITIONS. The derivative of the vector $\mathbf{A}(t)$ with respect to t is defined as the vector

$$\frac{d\mathbf{A}}{dt} = \operatorname*{Lim}_{\Delta t \to 0} \frac{\mathbf{A}(t + \Delta t) - \mathbf{A}(t)}{\Delta t} \qquad (9\text{-}1)$$

provided this limit exists. (All further discussions will tacitly include the assumption of existence of all derivatives considered.) Observe that since $\mathbf{A}$ may be undergoing simultaneous changes in both magnitude and direction, $d\mathbf{A}/dt$ can be expected to have any magnitude and direction.

The geometric notions of direction and magnitude are devoid of meaning without a reference frame in which these are to be reckoned. We

7

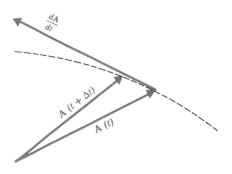

conceive a reference frame as a set of points in space, the distance between every pair of points remaining constant. At least four non-coplanar points are required. With the introduction of a suitable coordinate system in the reference frame, the two end points defining a vector may be located and a mathematical description of the vector given in terms of a set of components.

Consider the appearance of a vector **A** from two different reference frames, frame β rotating with respect to frame α. Suppose that the vector appears to be constant in frame β, so that the β-observed derivative of **A** is zero. Then, from frame α, **A** will appear to be changing its direction, so that the α-observed derivative of **A** is *not* zero. Thus, starting with a single vector, we can define more than one *different* vector by differentiation. The derivative of **A** depends on **A** *and* the reference frame from which its change is observed.

If reference frame β is translating, but not rotating, relative to frame α, the β-observed derivative of a vector will be equal to the α-observed derivative of the vector. This follows from the fact that the vector is defined only in terms of the relative positions of its end points in each reference frame, without regard for the position of the head of the vector. (See Section 3-1.)

Whenever there is more than one reference frame under consideration, we will denote the α-observed derivative of a vector as

$$\frac{d_\alpha \mathbf{A}}{dt} = \overset{\alpha}{\mathbf{A}} \tag{9-2}$$

When only one reference frame can be inferred from the context of the discussion, the derivative of the vector **A** in that reference frame will be denoted as

$$\frac{d\mathbf{A}}{dt} = \dot{\mathbf{A}} \tag{9-3}$$

In Newtonian dynamics, we will use this last notation to indicate derivatives observed from an inertial reference frame. The derivative of a scalar

p is independent of reference frame and will be denoted as

$$\frac{dp}{dt} = \dot{p} \tag{9-4}$$

The detailed analysis of kinematics relative to rotating reference frames will be taken up in Chapter 14.

9-2

DERIVATIVES OF SUMS AND PRODUCTS.   The derivative of $\mathbf{A} \pm \mathbf{B}$ is the vector

$$\frac{d}{dt}(\mathbf{A} \pm \mathbf{B}) = \lim_{\Delta t \to 0} \frac{[\mathbf{A}(t + \Delta t) \pm \mathbf{B}(t + \Delta t)] - [\mathbf{A}(t) \pm \mathbf{B}(t)]}{\Delta t}$$

$$= \lim_{\Delta t \to 0} \frac{\mathbf{A}(t + \Delta t) - \mathbf{A}(t)}{\Delta t} \pm \lim_{\Delta t \to 0} \frac{\mathbf{B}(t + \Delta t) - \mathbf{B}(t)}{\Delta t}$$

$$= \dot{\mathbf{A}} \pm \dot{\mathbf{B}} \tag{9-5}$$

Therefore differentiation of a sum or difference of vectors follows the same rule as differentiation of a sum or difference of scalars.

Differentiation of a product of a scalar and a vector may be carried out as follows.

$$\frac{d}{dt}[p(t)\,\mathbf{A}(t)] = \lim_{\Delta t \to 0} \frac{(p + \Delta p)(\mathbf{A} + \Delta \mathbf{A}) - p\mathbf{A}}{\Delta t}$$

$$= \lim_{\Delta t \to 0} \left(\frac{p\,\Delta \mathbf{A}}{\Delta t} + \frac{\Delta p \mathbf{A}}{\Delta t} + \frac{\Delta p\,\Delta \mathbf{A}}{\Delta t} \right)$$

$$= p\dot{\mathbf{A}} + \dot{p}\mathbf{A} \tag{9-6}$$

As with sums and differences, we have an analogy with products of scalars.

Similar considerations lead to the following formulas for derivatives of dot and cross products.

$$\frac{d}{dt}(\mathbf{A} \cdot \mathbf{B}) = \mathbf{A} \cdot \dot{\mathbf{B}} + \dot{\mathbf{A}} \cdot \mathbf{B} \tag{9-7}$$

$$\frac{d}{dt}(\mathbf{A} \times \mathbf{B}) = \mathbf{A} \times \dot{\mathbf{B}} + \dot{\mathbf{A}} \times \mathbf{B} \tag{9-8}$$

9-3

USEFUL RESOLUTIONS.   The derivative of a vector is another vector and may be resolved in a number of ways. Two of the most useful in dynamics are the following.

"Swinging" and "Stretching" Components. The vector **A** may be expressed in terms of a unit vector $\mathbf{u}_A$ in the direction of **A** as

$$\mathbf{A} = A\,\mathbf{u}_A$$

Application of (9-6) then gives us

$$\dot{\mathbf{A}} = A\dot{\mathbf{u}}_A + \dot{A}\mathbf{u}_A \tag{9-9}$$

The geometric significance of each term in this resolution is important and can be inferred from Figure 9-1. We see from this figure that since

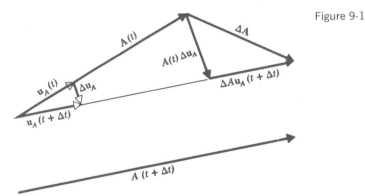

Figure 9-1

the magnitude of $\mathbf{u}_A$ is by definition fixed as unity, its incremental change $\Delta\mathbf{u}_A$ is approximately perpendicular to **A**. Similarly, the two components of $\Delta\mathbf{A}$, $A(t)\,\Delta\mathbf{u}_A$ and $\Delta A\,\mathbf{u}_A(t + \Delta t)$, are approximately perpendicular and parallel, respectively, to $\mathbf{A}(t)$. As $\Delta t \to 0$, the approximations become arbitrarily accurate, so that the two terms in the right-hand side of Equation 9-9 represent components perpendicular and parallel to **A**. The component $A\,\dot{\mathbf{u}}_A$ might be aptly termed the "swinging" component of $\dot{\mathbf{A}}$, and the component $\dot{A}\,\mathbf{u}_A$ termed the "stretching" component of $\dot{\mathbf{A}}$.

Rectangular Cartesian Components. If the unit vectors $\mathbf{u}_x$, $\mathbf{u}_y$, and $\mathbf{u}_z$ are defined to *remain fixed* in the reference frame from which the derivative is reckoned, their derivatives will be zero. In this circumstance, differentiation of the expansion

$$\mathbf{A} = A_x\mathbf{u}_x + A_y\mathbf{u}_y + A_z\mathbf{u}_z \tag{[3-16]*}$$

results in

$$\dot{\mathbf{A}} = \dot{A}_x\mathbf{u}_x + \dot{A}_y\mathbf{u}_y + \dot{A}_z\mathbf{u}_z \tag{9-10}$$

Equations 9-9 and 9-10 are two of the most useful ways of expressing

* Brackets signify an equation that is repeated from an earlier development.

the derivative of a vector. A thorough understanding of these resolutions of the derivative is essential to a successful study of kinematics.

Example

The rectangular Cartesian components of a vector $\mathbf{r}(t)$ are given by

$$\mathbf{r}(t) = 4.8 \, t \, \mathbf{u}_x + (7.2t - 4.9t^2)\mathbf{u}_z$$

Evaluate the rectangular Cartesian components of the derivative $\mathbf{v} = \dot{\mathbf{r}}$ and $\mathbf{a} = \dot{\mathbf{v}}$ and, for $t = 1$, evaluate the "stretching" and "swinging" components of these vectors.

The rectangular Cartesian components may be written using (9-10) as

$$\mathbf{v}(t) = \dot{\mathbf{r}} = 4.8 \, \mathbf{u}_x + (7.2 - 9.8t)\mathbf{u}_z$$
$$\mathbf{a} = \dot{\mathbf{v}} = -9.8 \, \mathbf{u}_z$$

At $t = 1$, these have the values

$$\mathbf{v}(1) = 4.8 \, \mathbf{u}_x - 2.6 \, \mathbf{u}_z$$
$$\mathbf{a} = -9.8 \, \mathbf{u}_z$$

The resultants of these vectors are shown in Figure 9-2, along with the path traced out by the head of the vector $\mathbf{r}(t)$ as it varies. Also shown are the "stretching" and "swinging" components, which may be evaluated as follows.

The component of $\dot{\mathbf{r}}$ in the direction of $\mathbf{r}$ has the magnitude

$$
\begin{aligned}
v_r &= v \cos \angle^{\mathbf{v}}\mathbf{r} \\
&= \mathbf{v} \cdot \mathbf{u}_r \\
&= \mathbf{v} \cdot \frac{\mathbf{r}}{r} \\
&= \frac{(4.8)(4.8) + (-2.6)(2.3)}{\sqrt{(4.8)^2 + (2.3)^2}} \\
&= 3.21
\end{aligned}
$$

The "swinging" component of $\mathbf{v}$ (not labeled) has the magnitude

$$
\begin{aligned}
\sqrt{v^2 - v_r^2} &= \sqrt{(4.8)^2 + (2.6)^2 - (3.21)^2} \\
&= 4.42
\end{aligned}
$$

We may similarly evaluate the component of $\mathbf{a}$ in the direction of $\mathbf{v}$ as

$$
\begin{aligned}
a_t &= \frac{\mathbf{a} \cdot \mathbf{v}}{v} \\
&= \frac{(0)(4.8) + (-9.8)(-2.6)}{\sqrt{(4.8)^2 + (2.6)^2}} \\
&= 4.67
\end{aligned}
$$

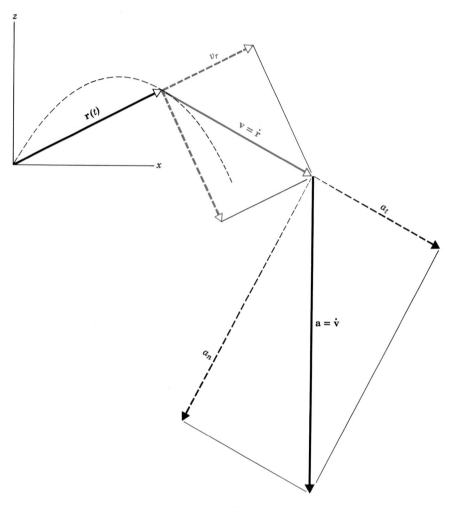

Figure 9-2

and the component perpendicular to **v** as

$$a_n = \sqrt{a^2 - a_t^2}$$
$$= \sqrt{(9.8)^2 - (4.67)^2}$$
$$= 8.62$$

Although the rectangular Cartesian components will define either of the vectors $\dot{\mathbf{r}}$ or $\dot{\mathbf{v}}$ as well as the "stretching" and "swinging" components, we find that these latter components greatly simplify a number of problems in kinematics.

Problems

9-1   If $\dot{A} = 0$, does it follow that $\dot{\mathbf{A}} = \mathbf{0}$? Explain. If $\mathbf{A} = \mathbf{0}$, does it follow that $\dot{A} = 0$?

9-2   The rectangular Cartesian components of a vector are

$$\mathbf{A} = t\mathbf{u}_x + (9t - 4t^2)\mathbf{u}_y$$

Evaluate $\dot{\mathbf{A}}, \dot{A}, |\dot{\mathbf{A}}|,$ and $\ddot{\mathbf{A}}$ at $t = \frac{9}{8}$ and at $t = \frac{5}{4}$. Show $\dot{\mathbf{A}}$ and $\ddot{\mathbf{A}}$ on a sketch.

9-3   For each of the following, evaluate $\dot{\mathbf{A}}, \ddot{\mathbf{A}}, |\dot{\mathbf{A}}|, \dot{A},$ and $\ddot{A},$ at $t = 0$.
(a) $\mathbf{A} = 5t^2\mathbf{u}_x + 2t\,\mathbf{u}_y - t^3\,\mathbf{u}_z$.
(b) $\mathbf{A} = \cos 2t\,\mathbf{u}_x + \sin 2t\,\mathbf{u}_y + e^{-t/2}\,\mathbf{u}_z$.
(c) $\mathbf{A} = \cos \omega t\,\mathbf{u}_x + \sin \omega t\,\mathbf{u}_y + bt\,\mathbf{u}_z$.
(d) $\mathbf{A} = 2t\,\mathbf{u}_x + \log(1 + t^2)\mathbf{u}_y + \tan^{-1}(1 + t)\mathbf{u}_z$.

9-4   For $\mathbf{A} = t\,\mathbf{u}_x - t^2\mathbf{u}_y + (t - 2)\mathbf{u}_z$ and $\mathbf{B} = (3t - 1)\,\mathbf{u}_x + \mathbf{u}_y - t\mathbf{u}_z,$ evaluate $(d/dt)(\mathbf{A} \cdot \mathbf{B}),$ $(d/dt)(\mathbf{A} \times \mathbf{B}),$ $(d/dt)|\mathbf{A} + \mathbf{B}|,$ and $(d/dt)(\mathbf{A} \times \dot{\mathbf{B}}),$ at $t = 2$.

9-5   In terms of $\mathbf{A}, \mathbf{B},$ and their derivatives, evaluate $(d/dt)(\mathbf{A} \cdot \dot{\mathbf{B}} - \dot{\mathbf{A}} \cdot \mathbf{B})$.

9-6   Evaluate $(d/dt)[\mathbf{A} \times (\mathbf{B} \times \mathbf{C})],$ for
$$\mathbf{A} = \cos \omega t\,\mathbf{u}_x + \sin \omega t\,\mathbf{u}_y + \omega t\,\mathbf{u}_z$$
$$\mathbf{B} = -\sin \omega t\,\mathbf{u}_x + \cos \omega t\,\mathbf{u}_y + \omega\mathbf{u}_z$$
$$\mathbf{C} = \mathbf{u}_z$$

9-7   For the vector $\mathbf{A} = (t^2 - 4t)\mathbf{u}_x + (t^2 + 2t)\mathbf{u}_y + t(t^2 - 1)\mathbf{u}_z,$ evaluate the components of $\dot{\mathbf{A}}$ parallel to $\mathbf{A}$ and perpendicular to $\mathbf{A}$. (Suggestion. See Equation 3-15.)

9-8   For $\mathbf{r} = a \cos \omega t\mathbf{u}_x + a \sin \omega t\mathbf{u}_y + bt\mathbf{u}_z,$ sketch the curve traced out by the head of $\mathbf{r}$, where the tail remains at a fixed point 0. Sketch $\mathbf{v} = \dot{\mathbf{r}}$ and $\mathbf{a} = \ddot{\mathbf{r}}$ for some arbitrary value of t.

9-9   For the vector $\mathbf{r} = \rho(t) \cos \phi(t)\mathbf{u}_x + \rho(t) \sin \phi(t)\mathbf{u}_y + z(t)\mathbf{u}_z$:
(a) Evaluate $\dot{\mathbf{r}}$ and $\ddot{\mathbf{r}}$.
(b) Express $\mathbf{u}_x$ and $\mathbf{u}_y$ in terms of $\mathbf{u}_\rho$ and $\mathbf{u}_\phi$.
(c) Express $\mathbf{r}, \dot{\mathbf{r}},$ and $\ddot{\mathbf{r}}$ in terms of $\mathbf{u}_\rho, \mathbf{u}_\phi,$ and $\mathbf{u}_z$.

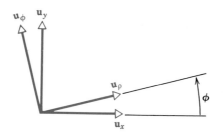

9-10   Verify Equation 9-7 using Equations 9-10 and 3-20.

9-11   Verify Equation 9-8 using Equations 9-10 and 3-22.

9-12   Show that $\mathbf{H} = e^{-\lambda x} (\mathbf{C}_1 \sin \lambda y + \mathbf{C}_2 \cos \lambda y)$ satisfies

$$\frac{\partial^2 \mathbf{H}}{\partial x^2} + \frac{\partial^2 \mathbf{H}}{\partial y^2} = \mathbf{0}$$

9-13   Show analytically that if $\ddot{\mathbf{A}}$ always remains parallel to $\mathbf{A}$, then the component of $\dot{\mathbf{A}}$ in the direction perpendicular to $\mathbf{A}$ has magnitude inversely proportional to the magnitude of $\mathbf{A}$.

9-14   A vector $\mathbf{r}$ varies with t in such a way that its magnitude remains constant. Show formally that this implies that $\mathbf{r} \cdot \dot{\mathbf{r}} = 0$. What does this mean geometrically?

9-15   Based on a sketch similar to Figure 9-1, show that $(d/dt)[p(t)\, \mathbf{A}(t)]$ $= p\, \dot{\mathbf{A}} + \dot{p}\, \mathbf{A}$, where A is not necessarily fixed. Do the two terms in the right-hand member represent components perpendicular and parallel to $\mathbf{A}$?

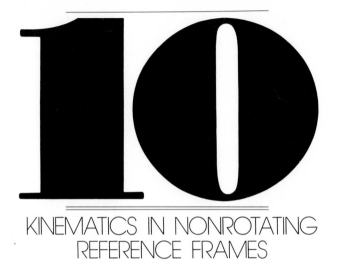

KINEMATICS IN NONROTATING REFERENCE FRAMES

Before we can begin a successful study of the interactions that affect the motions of bodies, we must analyze motions themselves. Such a study, without regard to the forces that may cause the motions, is called kinematics. As a beginning, motions of a particle (or, as far as kinematics is concerned, a point in space) will be considered. This will be followed by a study of motions of rigid bodies, which are more complicated. More complicated still is the kinematics of deformable objects, treatment of which will not be undertaken here.

10-1
POSITION, VELOCITY, AND ACCELERATION OF A PARTICLE. We often begin the analysis of the motion of a particle in a reference frame by specifying its position relative to some point fixed to the reference frame through the introduction of a *position vector*. If we know this position vector as a function of time, we can then determine its time rate of change, or the *velocity* of the particle in the reference frame. The time rate of change of this velocity vector, or the *acceleration* of the particle, will

15

be of equal importance to our study. The basic idea here is illustrated in Figure 10-1. Figure 10-1*a* shows the path of the particle traced out by the

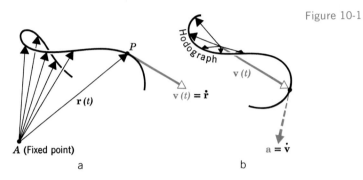

Figure 10-1

head of the position vector **r** as it changes with time. The "path" in Figure 10-1*b*, called the *hodograph* of the motion, is related to the changing velocity vector in the same way that the path of the particle is related to the changing position vector.

Note that since the velocity is related to the *change* in position, neither the velocity nor the acceleration depends on the location of the reference point for position.

One-Dimensional Motion.   If a point moves along a straight line, its position relative to a point fixed on the line may be specified by

$$\mathbf{r}(t) = x(t)\mathbf{u}_x \tag{10-1a}$$

where the x axis is defined to lie along the line of motion. The velocity and acceleration are then related to the distance x by

$$\mathbf{v}(t) = \dot{x}(t)\mathbf{u}_x \tag{10-1b}$$

$$\mathbf{a}(t) = \ddot{x}(t)\mathbf{u}_x \tag{10-1c}$$

Example

A particle is fired vertically upward with an initial speed v_0 and accelerates downward at the uniform rate g. Determine the height as a function of time and the maximum height attained.

Let the z axis be defined positive upward. Then the acceleration is given by

$$\mathbf{a} = -g\mathbf{u}_z$$

Since this is the derivative of the velocity,

$$\mathbf{v} = (-gt + C)\mathbf{u}_z$$

Now, if we count time starting with the instant at which the velocity is $v_0\mathbf{u}_z$, the value of the constant of integration must be $C = v_0$. The velocity is then

$$\mathbf{v} = (v_0 - gt)\mathbf{u}_z$$

Since this is the derivative of the position vector $\mathbf{r}$,

$$\mathbf{r} = (z_0 + v_0t - \tfrac{1}{2}gt^2)\mathbf{u}_z$$

The constant of integration z_0 gives the position of the particle at time $t = 0$ and depends on the choice for the origin of the z axis. Figure 10-2

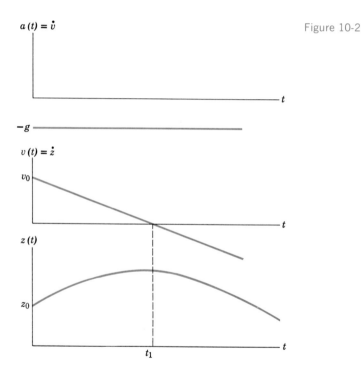

Figure 10-2

shows how the position, velocity, and acceleration vary with time in this example. We see from these graphs that the maximum height is reached at the time when $v = 0$, or $t_1 = v_0/g$. Substitution of this value into the expression for $\mathbf{r}$ gives the maximum height as the magnitude of the position vector.

$$\mathbf{r}_1 = \left(z_0 + \frac{v_0^2}{2g}\right)\mathbf{u}_z$$

Example

In the Scotch yoke mechanism shown in Figure 10-3, the rod is driven back and forth by pin B as it moves in the circular path at constant speed

Figure 10-3

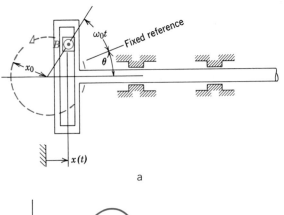

a

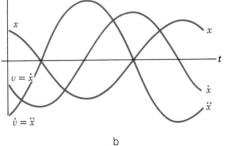

b

ω_0 (radions per unit time). Determine the velocity and acceleration of the rod.

We see from the figure that the position is given by

$$x = x_0 \cos (\omega_0 t + \theta)$$

The velocity of the rod is then given by

$$\mathbf{v} = \dot{x}\, \mathbf{u}_x$$
$$= -\omega_0 x_0 \sin (\omega_0 t + \theta)\mathbf{u}_x$$

and the acceleration by

$$\mathbf{a} = \dot{v}\, \mathbf{u}_x$$
$$= -\omega_0^2 x_0 \cos (\omega_0 t + \theta)\mathbf{u}_x$$

These quantities are illustrated in Figure 10-3*b*.

Observe that in this motion the displacement and acceleration are related by

$$\dot{v} = -\omega_0^2 x$$

or

$$\ddot{x} + \omega_0^2 x = 0$$

Models for a remarkable variety of physical systems lead to this differential equation. Its solution, $x(t)$, given above, is therefore important in a number of applications.

Periodic Motion.   The preceding is an example of *periodic* motion, so called because after some period of time, the motion repeats itself. In the special case in which the motion quantities vary according to trigonometric functions, as in the preceding example, the motion is called *sinusoidal*, or *simple harmonic*.

The *period p* of the motion is defined as the smallest time interval after which the motion repeats itself. In the preceding example, the period is

$$p = \frac{2\pi}{\omega_0}$$

The *frequency f* of a periodic motion is defined as the reciprocal of the period, or the number of cycles executed per unit of time. It is usually measured in hertz (Hz), defined as cycles per second. For the preceding example, the frequency would be

$$f = \frac{\omega_0}{2\pi}$$

The quantity ω_0 is properly called the *angular frequency* but, unfortunately, both ω_0 and f are often referred to as the frequency.

The *amplitude* of a periodic function is defined as the maximum absolute value that the function achieves. In the preceding example, the amplitude is of the displacement x_0.

Motion in a Circular Path.   Consider the motion of a point moving in a circular path, as shown in Figure 10-4. The dependence of the velocity and acceleration on the radius and the angle $\phi(t)$ will be of interest in a number of applications.

During a time increment Δt, the position vector undergoes a change given by the approximation

$$\Delta\mathbf{r} \approx \rho\,\Delta\phi\mathbf{u}_\phi$$

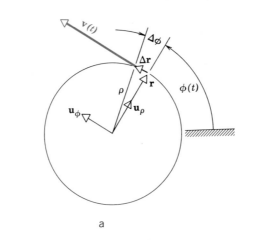

Figure 10-4

a

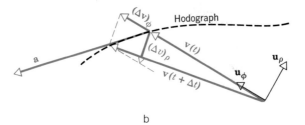

b

As $\Delta t \to 0$, the approximation

$$\mathbf{v} \approx \frac{\Delta \mathbf{r}}{\Delta t} \approx \rho \frac{\Delta \phi}{\Delta t} \, \mathbf{u}_\phi$$

becomes arbitrarily accurate, and we have

$$\mathbf{v} = \rho \, \dot{\phi} \, \mathbf{u}_\phi \tag{10-2}$$

The acceleration $\mathbf{a} = d\mathbf{v}/dt$ has a component perpendicular to $\mathbf{v}$, which may be computed in the same way in which $d\mathbf{r}/dt$ was just evaluated. Referring to Figure 10-4b, we see that

$$(\Delta \mathbf{v})_\rho \approx -v \, \Delta\phi \mathbf{u}_\rho$$

or,

$$\mathbf{a}_\rho = -v \, \dot{\phi} \, \mathbf{u}_\rho$$

Substitution from (10-2) leads to the alternate forms

$$\mathbf{a}_\rho = -\frac{v^2}{\rho} \, \mathbf{u}_\rho$$

$$\mathbf{a}_\rho = -\rho \, \dot{\phi}^2 \, \mathbf{u}_\rho$$

The minus sign expresses the fact that this radial component $\mathbf{a}_\rho$ is directed toward the center of the circle. From the same figure we see that the component of velocity change in the direction of $\mathbf{v}$ is given by the approximation

$$(\Delta\mathbf{v})_\phi \approx \Delta|\mathbf{v}|\; \mathbf{u}_\phi$$

which leads to

$$\mathbf{a}_\phi = \dot{v}\, \mathbf{u}_\phi$$

This may also be expressed in terms of the angle ϕ by introducing (10-2).

$$\mathbf{a}_\phi = \rho\, \ddot{\phi}\, \mathbf{u}_\phi$$

These two components of the acceleration,

$$\mathbf{a} = \mathbf{a}_\rho + \mathbf{a}_\phi$$

$$= -\frac{v^2}{\rho}\, \mathbf{u}_\rho + \dot{v}\, \mathbf{u}_\phi \tag{10-3a}$$

$$= -\rho\, \dot{\phi}^2\, \mathbf{u}_\rho + \rho\, \ddot{\phi}\, \mathbf{u}_\phi \tag{10-3b}$$

are the same "stretching" and "swinging" components that were discussed in Section 9-3 for the vector $\mathbf{A}(t)$.

Example

Evaluate the acceleration of a satellite traveling at a constant speed of 7617 m/s in a circular orbit of radius 6880 km.

Since the speed is constant, $\dot{v} = 0$, and Equation 10-3a yields the magnitude of the acceleration directed toward the earth as

$$a_\rho = \frac{(7617\ \text{m/s})^2}{6.88 \times 10^6\,\text{m}}$$

$$= 8.43\ \text{m/s}^2$$

Intrinsic Components of Velocity and Acceleration. Consider a particle P moving along a smooth path through space, as shown in Figure 10-5. We define a unit vector $\mathbf{u}_t$ to be tangent to the path at the location of P. In terms of this, the velocity of the particle is then given in terms of its speed v as

$$\mathbf{v} = v\, \mathbf{u}_t \tag{10-4}$$

Application of Equation 9-9 results in the following expression for the acceleration.

$$\dot{\mathbf{v}} = v\, \dot{\mathbf{u}}_t + \dot{v}\, \mathbf{u}_t$$

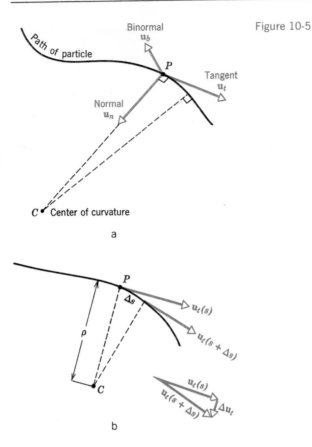

Figure 10-5

The second term in this expansion represents the acceleration component tangent to the path of the particle. The first term, representing the acceleration component normal to the path, needs further investigation. Using the chain rule, we may write this term in terms of the increment of arc length, ds, along the path.

$$v\dot{\mathbf{u}}_t = v\,\frac{ds}{dt}\,\frac{d\mathbf{u}_t}{ds} = v^2\,\frac{d\mathbf{u}_t}{ds}$$

The vector $d\mathbf{u}_t/ds$ defines the direction of the unit normal vector $\mathbf{u}_n$.

$$\frac{d\mathbf{u}_t}{ds} = \kappa\,\mathbf{u}_n \tag{10-5a}$$

The magnitude $\kappa = |d\mathbf{u}_t/ds|$ is called the *curvature* of the path. Its significance can be clarified by examining Figure 10-5b, which shows the plane containing $\mathbf{u}_t$ and $d\mathbf{u}_t/ds$. As $\Delta s \to 0$, the intersection of the lines normal to the path, from the positions s and $s + \Delta s$, approaches the center of curvature of the path. An inspection of the two similar triangles reveals that

$$\frac{|\Delta \mathbf{u}_t|}{\Delta s} \approx \frac{|\mathbf{u}_t|}{\rho} = \frac{1}{\rho}$$

where ρ is the local radius of curvature of the path. As $\Delta s \to 0$, this becomes $|d\mathbf{u}_t/ds| = 1/\rho$, so that the curvature κ is the reciprocal of the radius of curvature.

$$\kappa = \frac{1}{\rho} \qquad (10\text{-}5\mathrm{b})$$

With these results the above expression for acceleration may be written in terms of components tangential and normal to the path as

$$\mathbf{a} = \dot{v}\mathbf{u}_t + \frac{v^2}{\rho}\,\mathbf{u}_n \qquad (10\text{-}6)$$

Comparison of this result with Equation 10-3a reveals that the path may be considered as locally circular, the plane and radius of the equivalent circle varying along the path.

Example

A particle launched at speed $v_0 = 8.65$ m/s in a direction of 56.3° from the horizontal will, under the influence of the earth's gravity, follow the path shown in Figure 9-2. (This may be verified by procedures to be discussed in the next chapter.) With t as time in seconds and r in meters, the vector $\mathbf{r}(t)$ of the example on pp. 11–12 will give the position of the particle as it moves. What will be the radius of curvature of the path at the location of the particle at $t = 1$ s?

By its definition, the component a_n, determined earlier, is the component of $\mathbf{a}$ perpendicular to the path of the particle. But, from Equation 10-6, this component may also be expressed in terms of the speed and the radius of curvature we seek.

$$a_n = \frac{v^2}{\rho}$$

Therefore the radius of curvature may be determined from quantities evaluated in the earlier example, as

$$\rho = \frac{v^2}{a_n}$$

$$= \frac{(4.8 \text{ m/s})^2 + (-2.6 \text{ m/s})^2}{8.62 \text{ m/s}^2}$$

$$= 3.46 \text{ m}$$

Example

A particle is constrained to move along the helical path shown in Figure 10-6. Its speed varies according to $v = v_0 + bt$. Determine the acceleration

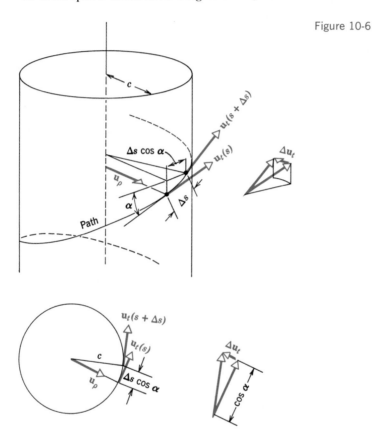

Figure 10-6

of the particle in terms of the initial speed v_0, the constant rate of change of speed b, the radius of the cylinder c, and the helix angle α.

As the particle moves through a distance Δs along the path, the local tangent vector changes direction, as shown in Figure 10-6. The projection of $\mathbf{u}_t$ onto the plane perpendicular to the cylinder axis has a length equal to $|\mathbf{u}_t| \cos \alpha = \cos \alpha$, as indicated in the lower diagram. Because $\Delta \mathbf{u}_t$ is parallel to this plane, its projection onto the plane is equal to $\Delta \mathbf{u}_t$. Now, considering the similarity of the vector triangle formed by these projections and the corresponding sector of the circle, we obtain

$$\frac{|\Delta \mathbf{u}_t|}{\Delta s \cos \alpha} = \frac{\cos \alpha}{c}$$

This and the observation that $\Delta \mathbf{u}_t$ is directed opposite to $\mathbf{u}_\rho$ lead to

$$\frac{d\mathbf{u}_t}{ds} = -\frac{\cos^2 \alpha}{c} \mathbf{u}_\rho$$

From Equations 10-5 and 10-6, the acceleration may now be evaluated as

$$\mathbf{a} = -\frac{v^2 \cos^2 \alpha}{c} \mathbf{u}_\rho + \dot{v} \, \mathbf{u}_t$$

$$= -\frac{(v_0 + bt)^2 \cos^2 \alpha}{c} \mathbf{u}_\rho + b \, \mathbf{u}_t$$

This result contains, as a special case, the acceleration of a point moving in a plane circular path. Does it agree with Equation 10-3a? What happens when $\alpha = \pi/2$?

Special Coordinate Systems. The equations that govern a physical system take on forms that vary with the choice of the coordinate system introduced. Whether useful information can be obtained from the equations often depends on whether the proper choice has been made. Problems in later chapters will furnish elucidation for this point. In this section we develop the necessary expressions for velocity and acceleration of a moving point, in three commonly useful coordinate systems.

Rectangular Cartesian coordinates. Perhaps the most commonly used coordinates for locating a moving particle are the rectangular Cartesian coordinates $x(t)$, $y(t)$, and $z(t)$. In terms of these, the position of a point is specified by

$$\mathbf{r} = x \, \mathbf{u}_x + y \, \mathbf{u}_y + z \, \mathbf{u}_z$$

The velocity and acceleration, observed from the reference frame to which the axes are fixed, are then expressed as

$$\mathbf{v} = \dot{x} \, \mathbf{u}_x + \dot{y} \, \mathbf{u}_y + \dot{z} \, \mathbf{u}_z$$

$$\mathbf{a} = \ddot{x} \, \mathbf{u}_x + \ddot{y} \, \mathbf{u}_y + \ddot{z} \, \mathbf{u}_z$$

A reexamination of the example on pp. 11 and 12 will yield some insight into these relationships and their connection with the intrinsic components of velocity and acceleration.

Circular cylindrical coordinates. For certain applications the cylindrical coordinates shown in Figure 10-7 prove to be the most suitable. These coordinates, ρ, ϕ, and z, are related to the rectangular Cartesian coordinates through the transformation

$$x = \rho \cos \phi$$

$$y = \rho \sin \phi$$

$$z = z$$

Components of displacement, velocity, and acceleration are resolved into the directions defined by the mutually perpendicular unit vectors $\mathbf{u}_\rho$, $\mathbf{u}_\phi$, and $\mathbf{u}_z$, shown in Figure 10-7. Because the directions of $\mathbf{u}_\rho$ and $\mathbf{u}_\phi$ are

Figure 10-7

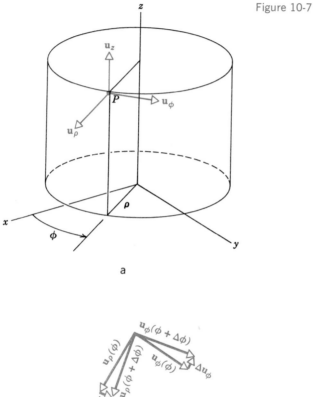

a

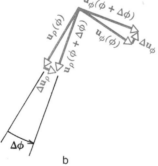

b

defined by the location of the moving point P, they vary with ϕ in the manner indicated in Figure 10-7b.

$$\frac{d\mathbf{u}_\rho}{d\phi} = \mathbf{u}_\phi \qquad \frac{d\mathbf{u}_\phi}{d\phi} = -\mathbf{u}_\rho$$

Or, with use of the chain rule,

$$\dot{\mathbf{u}}_\rho = \dot{\phi}\, \mathbf{u}_\phi \qquad \dot{\mathbf{u}}_\phi = -\dot{\phi}\, \mathbf{u}_\rho \tag{a}$$

From the resolution of the position vector,

$$\mathbf{r} = \rho\,\mathbf{u}_\rho + z\,\mathbf{u}_z \qquad (10\text{-}7a)$$

the velocity may be determined by differentiation and substitution of Equation a.

$$
\begin{aligned}
\mathbf{v} &= \dot{\mathbf{r}} \\
&= \dot\rho\,\mathbf{u}_\rho + \rho\,\dot{\mathbf{u}}_\rho + \dot z\,\mathbf{u}_z \\
&= \dot\rho\,\mathbf{u}_\rho + \rho\,\dot\phi\,\mathbf{u}_\phi + \dot z\,\mathbf{u}_z \qquad (10\text{-}7b)
\end{aligned}
$$

A second differentiation and use of Equation a give the acceleration as

$$
\begin{aligned}
\mathbf{a} &= \dot{\mathbf{v}} \\
&= \ddot\rho\,\mathbf{u}_\rho + \dot\rho\,\dot{\mathbf{u}}_\rho + \dot\rho\dot\phi\,\mathbf{u}_\phi + \rho\ddot\phi\,\mathbf{u}_\phi + \rho\dot\phi\,\dot{\mathbf{u}}_\phi + \ddot z\,\mathbf{u}_z \\
&= (\ddot\rho - \rho\dot\phi^2)\mathbf{u}_\rho + (\rho\ddot\phi + 2\dot\rho\dot\phi)\mathbf{u}_\phi + \ddot z\,\mathbf{u}_z \qquad (10\text{-}7c)
\end{aligned}
$$

Example

Determine the acceleration of the particle following the helical path shown in Figure 10-6. The speed of the particle varies according to $v = v_0 + bt$.

The position at time t is given by

$$\rho = c$$

$$\phi = \frac{s\cos\alpha}{c}$$

$$z = s\sin\alpha$$

in which s is the arc length, given by

$$s = s_0 + v_0 t + \tfrac{1}{2}bt^2$$

The velocity may be determined by substitution of these values into (10-7b).

$$
\begin{aligned}
\mathbf{v} &= \dot s\cos\alpha\,\mathbf{u}_\phi + \dot s\sin\alpha\,\mathbf{u}_z \\
&= (v_0 + bt)\cos\alpha\,\mathbf{u}_\phi + (v_0 + bt)\sin\alpha\,\mathbf{u}_z
\end{aligned}
$$

The acceleration may be determined by substitution into (10-7c).

$$
\begin{aligned}
\mathbf{a} &= -c\dot\phi^2\,\mathbf{u}_\rho + c\ddot\phi\,\mathbf{u}_\phi + \ddot z\,\mathbf{u}_z \\
&= -c\left(\frac{s\cos\alpha}{c}\right)^2\mathbf{u}_\rho + c\,\frac{\ddot s\cos\alpha}{c}\,\mathbf{u}_\phi + \ddot s\sin\alpha\,\mathbf{u}_z \\
&= -\frac{v^2\cos^2\alpha}{c}\,\mathbf{u}_\rho + b\cos\alpha\,\mathbf{u}_\phi + b\sin\alpha\,\mathbf{u}_z
\end{aligned}
$$

This may be brought into the form expressed earlier in intrinsic components by making use of the relationship for $\mathbf{u}_t$ in terms of $\mathbf{u}_\phi$ and $\mathbf{u}_z$.

Spherical coordinates.   Another useful coordinate system makes use of r, ϕ, and θ to locate a moving particle, as indicated in Figure 10-8.* These

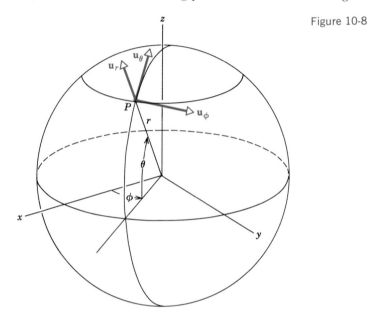

Figure 10-8

coordinates are related to the rectangular Cartesian coordinates through the transformation

$$x = r \cos \theta \cos \phi$$
$$y = r \cos \theta \sin \phi$$
$$z = r \sin \theta$$

Components of displacement, velocity, and acceleration are resolved into the directions defined by the mutually perpendicular unit vectors $\mathbf{u}_r$, $\mathbf{u}_\phi$, and $\mathbf{u}_\theta$, shown in Figure 10-8. Because the directions of all these are defined by the location of the moving point P, they vary with both ϕ and θ. As Problem 10-31 requires showing, these variations are expressed by

$$\frac{\partial \mathbf{u}_r}{\partial \phi} = \cos \theta \; \mathbf{u}_\phi \qquad\qquad \frac{\partial \mathbf{u}_r}{\partial \theta} = \mathbf{u}_\theta$$

$$\frac{\partial \mathbf{u}_\phi}{\partial \phi} = -\cos \theta \; \mathbf{u}_r + \sin \theta \; \mathbf{u}_\theta \qquad\qquad \frac{\partial \mathbf{u}_\phi}{\partial \theta} = \mathbf{0}$$

$$\frac{\partial \mathbf{u}_\theta}{\partial \phi} = -\sin \theta \; \mathbf{u}_\phi \qquad\qquad \frac{\partial \mathbf{u}_\theta}{\partial \theta} = -\mathbf{u}_r$$

* A slightly different spherical coordinate scheme is often introduced with the complement of θ instead of θ.

Differentiation of the position relationship,

$$\mathbf{r} = r\mathbf{u}_r \qquad (10\text{-}8a)$$

and use of the above expressions for the rates of change of the position vectors lead to the following expressions for velocity and acceleration in this spherical coordinate system.

$$\mathbf{v} = \dot{r}\,\mathbf{u}_r + r\cos\theta\,\dot{\phi}\,\mathbf{u}_\phi + r\,\dot{\theta}\,\mathbf{u}_\theta \qquad (10\text{-}8b)$$

$$\begin{aligned}
\mathbf{a} = {} & (\ddot{r} - r\cos^2\theta\,\dot{\phi}^2 - r\,\dot{\theta}^2)\,\mathbf{u}_r \\
& + (r\cos\theta\,\ddot{\phi} + 2\dot{r}\cos\theta\,\dot{\phi} - 2r\sin\theta\,\dot{\phi}\,\dot{\theta})\,\mathbf{u}_\phi \\
& + (r\ddot{\theta} + r\cos\theta\sin\theta\,\dot{\phi}^2 + 2\dot{r}\dot{\theta})\,\mathbf{u}_\theta \qquad (10\text{-}8c)
\end{aligned}$$

Velocity and Acceleration Differences. Kinematics analysis is sometimes simplified by considering the velocity or acceleration of a point in terms of those of a second point and the differences between those of the two points.

As an example, let us determine the velocity and acceleration of the man being reeled into the rescue helicopter shown in Figure 10-9. The line suspension point B, fixed to the helicopter, has the velocity and acceleration shown. The take-up of the line is specified by $\rho(t)$, and the pendulumlike swinging by $\phi(t)$.

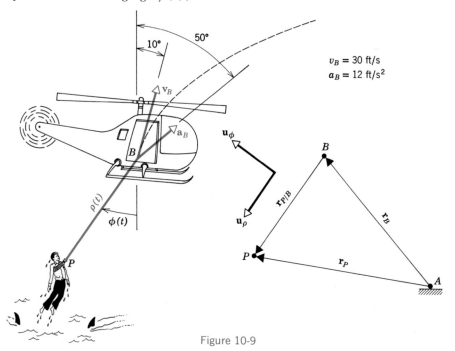

Figure 10-9

Introducing the position vectors locating the points B and P, relative to a fixed point A, we see that the location of point P may be expressed by

$$\boxed{\mathbf{r}_P = \mathbf{r}_B + \mathbf{r}_{P/B}} \qquad (10\text{-}9a)$$

Differentiation of this equation results in the expression

$$\dot{\mathbf{r}}_P = \dot{\mathbf{r}}_B + \dot{\mathbf{r}}_{P/B}$$

$$\boxed{\mathbf{v}_P = \mathbf{v}_B + \mathbf{v}_{P/B}} \qquad (10\text{-}9b)$$

for the velocity, and the expression

$$\dot{\mathbf{v}}_P = \dot{\mathbf{v}}_B + \dot{\mathbf{v}}_{P/B}$$

$$\boxed{\mathbf{a}_P = \mathbf{a}_B + \mathbf{a}_{P/B}} \qquad (10\text{-}9c)$$

for the acceleration. The velocity and acceleration differences,

$$\mathbf{v}_{P/B} = \mathbf{v}_P - \mathbf{v}_B$$

$$\mathbf{a}_{P/B} = \mathbf{a}_P - \mathbf{a}_B$$

may be determined in this case with reference to Equations 10-7.

$$\mathbf{r}_{P/B} = \rho\,\mathbf{u}_\rho$$

$$\mathbf{v}_{P/B} = \dot{\rho}\,\mathbf{u}_\rho + \rho\,\dot{\phi}\,\mathbf{u}_\phi$$

$$\mathbf{a}_{P/B} = (\ddot{\rho} - \rho\dot{\phi}^2)\mathbf{u}_\rho + (\rho\ddot{\phi} + 2\dot{\rho}\dot{\phi})\mathbf{u}_\phi$$

Referred to the $\mathbf{u}_\rho$ and $\mathbf{u}_\phi$ directions, the velocity and acceleration of point B are

$$\mathbf{v}_B = 30 \text{ ft/s } [-\cos(\phi - 10°)\mathbf{u}_\rho + \sin(\phi - 10°)\mathbf{u}_\phi]$$

$$\mathbf{a}_B = 12 \text{ ft/s}^2[-\cos(\phi - 50°)\mathbf{u}_\rho + \sin(\phi - 50°)\mathbf{u}_\phi]$$

The resultant velocity and acceleration point P can now be evaluated by substitution into Equations 10-9.

$$\mathbf{v}_P = [-30 \text{ ft/s } \cos(\phi - 10°) + \dot{\rho}]\,\mathbf{u}_\rho$$
$$+ [30 \text{ ft/s } \sin(\phi - 10°) + \rho\dot{\phi}]\,\mathbf{u}_\phi$$

$$\mathbf{a}_P = [-12 \text{ ft/s}^2 \cos(\phi - 50°) + \ddot{\rho} - \rho\dot{\phi}^2]\,\mathbf{u}_\rho$$
$$+ [12 \text{ ft/s}^2 \sin(\phi - 50°) + \rho\ddot{\phi} + 2\dot{\rho}\dot{\phi}]\,\mathbf{u}_\phi$$

It must be understood that the velocity and acceleration differences determined above are those observed from the helicopter *only if the helicopter is not rotating*. The derivatives that led from (10-9a) to (10-9b) to (10-9c) are those observed from the earth (or equally well from a nonrotating helicopter); as pointed out on p. 8, these derivatives are *not* equal to those observed from a rotating reference frame. For illustration, consider a motion in which, at some instant, the point B is stationary, the line length is fixed and the helicopter and line are rotating together at the rate $\dot{\phi}$. Since B is stationary, $\mathbf{v}_B = \mathbf{0}$ and, because of the rotation, $\mathbf{v}_P \neq \mathbf{0}$. Equation 10-9b then tells us that $\mathbf{v}_{P/B} \neq \mathbf{0}$. But, since the entire system is moving as a rigid body, the velocity of P would be zero as observed from the helicopter. Therefore $\mathbf{v}_{P/B}$ is *not* the velocity of P observed from the helicopter.

Detailed analysis of velocity and acceleration observed from rotating reference frames will be carried out in Chapter 14.

Problems

10-1   Under what circumstances are the expressions for distance and speed,

$$s = s_0 + v_0 t + \tfrac{1}{2}at^2$$
$$v = v_0 + at$$
$$v^2 = v_0^2 + 2a(s - s_0)$$

applicable? Under what circumstances does a in these formulas represent the acceleration?

10-2   Does a constant magnitude velocity imply a zero value of acceleration? Explain.

10-3   Determine the time taken for a particle with initial velocity $v(0) = v_0\,\mathbf{u}_z$ and constant acceleration $\mathbf{a} = -g\,\mathbf{u}_z$ to return to the point of this initial motion. What is its velocity at that time?

10-4   Starting from rest and traveling in a straight line, an automobile accelerates at a rate of 4 ft/s². Determine the time required for the car to reach a speed of 60 mi/h and the distance traveled in this interval.

10-5   An automobile is traveling at a speed of 60 km/h. The driver applies the brakes and decelerates at a rate of 4 m/s². Determine both the time and the distance required for the vehicle to come to rest.

10-6   The position of a point is given by the equation

$$\mathbf{r} = (1\text{ ft/s}^2)t^2\,\mathbf{u}_y + (2\text{ ft})e^{(1s^{-1})t}$$
$$+ (1\text{ ft}) \sin[(\pi\text{ s}^{-1})t]\,\mathbf{u}_z$$

Determine the position, velocity, and acceleration when $t = 0$ and when $t = 5$ s.

10-7 The velocity of a point is given by the equation

$$\mathbf{v} = (3 \text{ m/s})\mathbf{u}_x + [(1 \text{ m/s}^2)t + 2 \text{ m/s}]\mathbf{u}_y \\ + (2 \text{ m/s})e^{(1s^{-1})t} \mathbf{u}_z$$

At time $t = 1$ s, the position is $\mathbf{r}(1) = 5m\mathbf{u}_x + 2m\mathbf{u}_y + 1m\mathbf{u}_z$. Determine the general equations for the position and the acceleration at time t.

10-8 A particle moves with the simple harmonic motion in accordance with the equation

$$x = (10 \text{ ft}) \cos \left[\left(\frac{\pi}{2} \text{ s}^{-1} \right) t \right]$$

Determine the position, velocity, and acceleration at $t = 0$, 1 s, 2 s, 3 s, and 4 s. What is the period?

10-9 A point moves along a straight line with an acceleration $a = (2 \text{ ft/s}^3)t + 4 \text{ ft/s}^2$. At time $t = 5$ s, the velocity is 55 ft/s. What is the velocity at $t = 0$ and at $t = 10$ s? Determine the distance traveled from $t = 0$ to $t = 10$ s.

10-10 A point travels in a straight line with a constant acceleration of 8 ft/s². Determine the initial velocity, $v(0)$, if the point travels 500 m in 10 s.

10-11 Starting from rest, a point P travels in a straight line with an acceleration of 32 ft/s². A second point Q begins its motion 2 s later and somewhere along the path of point P. The point Q also starts from rest and travels with an acceleration of 32 ft/s². Point P overtakes point Q after point P travels 400 ft. Determine the distance, d, between the starting points of P and Q.

10-12 A point travels along a straight line from $x = 0$ to $x = 4$ ft. Determine the equation for the velocity and the time required to complete the motion when
(a) The acceleration is $a = 2 \text{ ft/s}^2 - (1 \text{ ft/s}^3)t$ and the initial velocity at $x = 0$ is 0.
(b) The acceleration is $a = (2 \text{ ft/s}^2)e^{(0.5 \text{ s}^{-1})t}$ and the initial velocity at $x = 0$ is 1.5 ft/s.

10-13 A point moves such that

$$\dot{x} = (2 \text{ m/s}^2)t \quad \text{and} \quad y = (1 \text{ m/s}^2)t^2 - 4 \text{ m}$$

At time $t = 0$, the value of x is 5 m. Determine the position, velocity, and acceleration when $t = 3$ s.

10-14 A point moves such that

$$\ddot{x} = 1 \text{ ft/s}^2 \quad \text{and} \quad y = (2 \text{ ft/s}^2)t^2 + 3 \text{ ft}$$

At time $t = 1$ s, the value of $\dot{x}$ is 2 ft/s and the value of x is 1 ft. Determine the position, velocity, and acceleration when $t = 5$ s.

10-15   In the absence of air resistance, an object hurled vertically near the surface of the earth is accelerated vertically according to the familiar $\ddot{z} = -g$ and, because of a peculiar effect of the earth's rotation called Coriolis acceleration (to be examined in Chapter 14), is also accelerated in the east-west direction according to

$$\ddot{x} = -(2\Omega \cos \theta)\dot{z}$$

in which x is the eastward displacement, Ω is the rotation rate of the earth (radians per unit time), and θ is the latitude. A rifle bullet is fired exactly vertically in your locality at 1000 m/s. If aerodynamic forces were negligible, where would it land?

10-16   The motion of a point along logarithmic spiral is specified in cylindrical coordinates as follows.

$$\rho = e^{\alpha t} \qquad \phi = \omega t$$

Express the position, velocity, and acceleration at $t = 0$ in terms of components on the $\mathbf{u}_\rho$ and $\mathbf{u}_\phi$ directions.

10-17   A point moves along a curved path such that its cylindrical coordinates are

$$\rho = a \sin \omega t \qquad \dot{\phi} = \omega \cos \omega t$$

At $t = 0$, the value of ϕ is 0. Express the position, velocity, and acceleration at $\omega t = \pi/4$ in terms of components in the $\mathbf{u}_\rho$ and $\mathbf{u}_\phi$ directions.

10-18   A point moves along a curved path such that its cylindrical coordinates are

$$\rho = a \sin^2 \omega t \qquad \phi = e^{\alpha t} - 1$$

Express the position, velocity, and acceleration at $t = 0$ in terms of components in the $\mathbf{u}_\rho$ and $\mathbf{u}_\phi$ directions.

10-19   The velocity and acceleration of a point are known to be

$$\mathbf{v} = (-8 \text{ ft/s})\mathbf{u}_x - (6 \text{ ft/s})\mathbf{u}_y \quad \text{and} \quad \mathbf{a} = (-4 \text{ ft/s}^2)\mathbf{u}_x + (20 \text{ ft/s}^2)\mathbf{u}_y.$$

Determine the normal and tangential components of acceleration. What is the radius of curvature of the path of motion?

10-20   A point moves along the path $y = (2 \text{ ft}) \log(x/1 \text{ ft})$ with a constant speed of 10 ft/s. Determine the normal and tangential components of acceleration at:
(a) $x = 1$ ft.
(b) $x = 10$ ft.

10-21   A point moves along the path $\rho = (1 \text{ ft}) \, e^{2\phi}$. At $\phi = 0$, the speed is 10 ft/s and increasing at the rate of 20 ft/s². Determine the normal and tangential components of acceleration.

10-22   A particle moves according to $\mathbf{v} = C_1 \, \mathbf{u}_x + (C_2 - gt)\mathbf{u}_y$, where C_1, C_2, and g are constants. Sketch the hodograph.

10-23   Estimate the velocity and acceleration of the earth as it travels around the sun. Assume a circular orbit of a radius of 92 600 000 mi.

10-24   The moon makes one complete revolution about the earth every 27.3 days. Assume that the moon travels in a circular path with a radius of 386 Mm and that the magnitude of the velocity is constant. Determine the magnitude of the normal component of acceleration of the moon.

10-25   A point P moves along the circumference of a circle of radius 10 in. The radius vector rotates at a constant rate of 1760 rpm. Determine the velocity and acceleration of P.

10-26   A point moves along the circumference of a circle of radius 4 ft. At $\phi = 3\pi/4$, the magnitude of the velocity is 8 ft/s. The magnitude of the velocity is decreasing at a rate of 12 ft/s². Determine the normal and tangential components of acceleration.

10-27   For a particle moving along a helical path, verify that the expressions for velocity and acceleration obtained using intrinsic components are consistent with those obtained using circular cylinder coordinates.

10-28   Show that for plane motion the magnitudes of the normal and tangential components of acceleration may be expressed in terms of rectangular coordinates as follows.

$$a_t = \dot{v} = \frac{\dot{x}\ddot{x} + \dot{y}\ddot{y}}{\sqrt{\dot{x}^2 + \dot{y}^2}}$$

$$a_n = \frac{v^2}{\rho} = \frac{|\dot{x}\ddot{y} - \dot{y}\ddot{x}|}{\sqrt{\dot{x}^2 + \dot{y}^2}}$$

10-29   The position of a point is given by the equation

$$\mathbf{r} = at^2 \, \mathbf{u}_x + be^{\alpha t} \, \mathbf{u}_y + c \sin \omega t \, \mathbf{u}_z$$

Determine the normal and tangential components of the acceleration at time $t = 0$. What is the radius of curvature of the path?

10-30 The equation for a conical helix is

$$\rho = at$$
$$\phi = \omega t$$
$$z = bt$$

For $a = 2$ m/s, $b = 1$ m/s, and $\omega = 1$ s^{-1}, determine the normal and tangential components of the acceleration of a point at time $t = \pi/2\omega$. What is the radius of curvature of the path?

10-31 Using geometric arguments similar to those indicated in Figure 10-7 for circular cylindrical coordinates, verify the expressions on p. 28 for the rates of change of the unit base vectors for spherical coordinates.

10-32 A wheel of radius a rolls at constant rate without slipping along a straight horizontal surface, its center moving at speed $v(t)$.
   (a) Write the equations that give the x and y coordinates of a point P on the rim of the wheel.
   (b) From the result in (a), determine the velocity and acceleration of P when it is in contact with the surface and when it is directly above the contact point.
   (c) Evaluate these velocities and accelerations using (10-2), (10-3), and (10-9).

10-33 Sketch the hodograph of a point on the rim of a wheel rolling at constant speed. What happens to the hodograph if the wheel is accelerating?

10-34 The wheel rolls without slip, its center moving at speed $v(t)$.

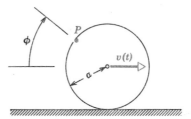

Evaluate the velocity and acceleration of the point P, which is attached to the wheel.

10-35 Let us analyze the motion of a particle following a circular path in another way.

(a) By formal differentiation of the position vector.

$$\mathbf{r} = \rho \cos \phi \; \mathbf{u}_x + \rho \sin \phi \; \mathbf{u}_y$$

show that the velocity and acceleration are given by

$$\mathbf{v} = -\rho \, \dot{\phi} \, \sin \phi \; \mathbf{u}_x + \rho \, \dot{\phi} \, \cos \phi \; \mathbf{u}_y$$

$$\mathbf{a} = (-\rho \, \ddot{\phi} \, \sin \phi - \rho \, \dot{\phi}^2 \cos \phi)\mathbf{u}_x + (\rho \, \ddot{\phi} \, \cos \phi - \rho \, \dot{\phi}^2 \sin \phi)\mathbf{u}_y$$

(b) Express the unit vectors $\mathbf{u}_x$ and $\mathbf{u}_y$ in terms of the unit vectors $\mathbf{u}_\rho$ and $\mathbf{u}_\phi$, and ϕ. Substituting this result into the results in (a), verify (10-2) and (10-3b).

10-36 An airplane makes a diving turn on a path of projected horizontal radius of 150 m and angle with the horizontal of 30° while accelerating along the path at 10 m/s². Find the resultant acceleration of the airplane when its speed is 110 m/s. At what angle should the airplane be banked in order that the pilot would experience no lateral forces?

10-37 A man climbs, with a constant velocity v, a vertical ladder fastened to the side of an upright cylinder whirling at angular velocity ω about its vertical axis. The cylinder has a radius of R and an angular acceleration of α. Determine the acceleration of the man and express it in terms of cylindrical coordinate unit vectors.

10-38 The line in Figure 10-9 is 15 m long and is being reeled in at a uniform rate of 700 mm/s. What are the velocity and acceleration of P if:

(a) $\phi = 12.97°$ (constant).

(b) $\phi = 12.97°$, $\dot{\phi} = -0.4$ rad/s, $\ddot{\phi} = -0.0368$ rad/s².

10-39   At the instant shown, slider B has an acceleration to the left of 500 mm/s² and zero velocity. Determine the acceleration of slider A.

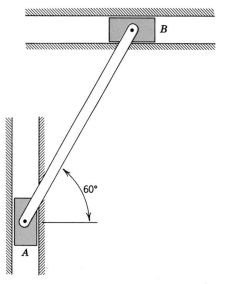

10-40   In the Scotch yoke mechanism of Figure 10-3, let point B move with the center of the driving block in the circular path, and let point P be fixed to the oscillating rod. Use Equations 10-9b and 10-9c to evaluate the velocity and acceleration of the rod.

10-41   Sliders A and B are free to rotate about the bars on which the slides travel. The bars are set at right angles to each other as shown. The velocity of B is a constant 10 ft/s. Determine the velocity and acceleration of A. (*Suggestion.* Note that $\mathbf{v}_{B/A}$ is perpendicular to $\mathbf{r}_{B/A}$.)

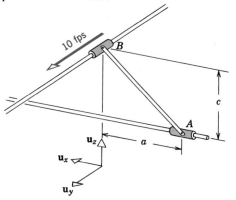

10-2

RIGID BODY MOTION. A motion in which all points on a rigid body have equal (not necessarily constant) velocities is called a *translational motion*. If the body is *rotating*, various points on it will have different velocities and accelerations. In this section an enlightening manner of characterizing translational and rotational motions is introduced and, from this, useful relationships among velocities and accelerations of various points are developed.

Rotation About an Axis. Consider the rigid body rotating about the fixed axis OC', as shown in Figure 10-10. The point P is moving in a

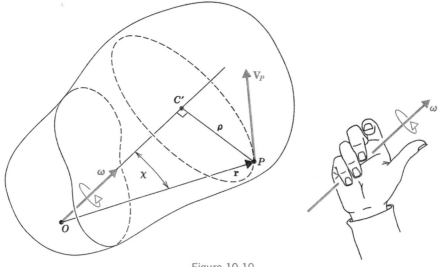

Figure 10-10

circular path in the plane perpendicular to the axis of rotation. We denote the rate of rotation (radians per unit time) as ω. Then, referring to Equation 10-7b, we see that the speed of point P is related to this angular rate by

$$v_P = \rho\dot{\phi} = \rho\omega \tag{a}$$

Now, if we define a position vector $\mathbf{r}$ from a point on the rotation axis to the point P, the distance ρ is given by

$$\rho = r\sin\chi \tag{b}$$

so that the speed can also be written as

$$v_P = r\omega\sin\chi \tag{c}$$

If we also define a vector $\boldsymbol{\omega}$ to have the magnitude ω, direction coinciding with the axis of rotation, and sense given by the direction of rotation according to the right hand rule, we may express the velocity of point P as

$$\boxed{\mathbf{v}_P = \boldsymbol{\omega} \times \mathbf{r}_P} \qquad (10\text{-}10)$$

The vector $\boldsymbol{\omega}$ is called the *angular velocity* of the body. It is of fundamental importance in the analysis of motions of rigid bodies.

Simultaneous Translation and Rotation. Consider next a somewhat more complicated motion of the body, in which the point Q is moving with velocity $\mathbf{v}_Q$ and the body is rotating about an axis through Q with angular velocity $\boldsymbol{\omega}$. Then the velocity difference $\mathbf{v}_{P/Q}$ is given by

$$\mathbf{v}_{P/Q} = \mathbf{v}_P - \mathbf{v}_Q$$
$$= \dot{\mathbf{r}}_P - \dot{\mathbf{r}}_Q$$
$$= \frac{d}{dt}(\mathbf{r}_P - \mathbf{r}_Q)$$
$$= \dot{\mathbf{r}}_{P/Q}$$
$$= \boldsymbol{\omega} \times \mathbf{r}_{P/Q}$$

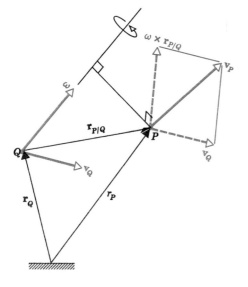

Or, when written as

$$\boxed{\mathbf{v}_P = \mathbf{v}_Q + \boldsymbol{\omega} \times \mathbf{r}_{P/Q}} \qquad (10\text{-}11)$$

the relationship expresses the motion of the body as a superposition of translation at the velocity of point Q and a rotation about an axis through Q.

Some insight into this relationship may be gained by examining some special, two-dimensional cases, in which the angular velocity is perpendicular to the velocities of all points on the body.

Example

The center of the wheel shown in Figure 10-11a is moving with speed v_C, while the wheel rotates at the rate ω. Figure 10-11b shows the velocities of

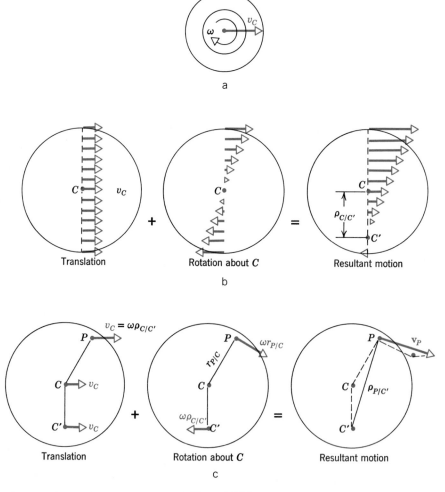

Figure 10-11

points along the diameter perpendicular to the direction of motion of the center.

Note the stationary point C', a distance $\rho_{C/C'} = v_{C/}\omega$ from the center of the wheel. This point is called the *instanteous center* for this motion. It is evident from Figure 10-11b that this point has, for an instant, the role of a center of rotation for points along the vertical diameter. It is also a center of rotation for every other point on the wheel, as we can learn from Figure 10-11c. The contribution to the velocity of point P, from rotation about C, has magnitude $\omega r_{P/C}$ and is perpendicular to CP. The contribution from translation has magnitude $v_C = \omega\rho_{C/C'}$ and is perpendicular to CC'. The triangle formed by these two components and their resultant is therefore similar to the triangle PCC', with ω as the constant of proportionality. This in turn implies that the resultant $\mathbf{v}_P$ has magnitude $\omega\rho_{P/C'}$ and is perpendicular to $C'P$. But this is exactly the velocity that results from a rotation about point C' at the rate ω! So it is quite reasonable to call the point C' the instantaneous center for this motion. Where would the instantaneous center be located if the wheel were rolling along a surface without slipping?

We see that the motion of the wheel, originally regarded as a superposition of a translation at the velocity of C and a rotation about an axis through C, may be viewed equally well as a pure rotation about the instantaneous center. Observe that the motion could also be viewed as a translation at the velocity of the arbitrary point P, together with a rotation about an axis through P. Note further that *the angular velocity involved in each of these resolutions of the motion is the same.*

Characterization of General Rigid Body Motion. The above observations lead one naturally to wonder whether, for every rigid body motion, there exists an angular velocity vector such that the motion can be described as a superposition of translation and rotation, as expressed by Equation 10-11 and, if so, whether there always exists an instantaneous axis of rotation, such that the motion may be described as simply a rotation, as expressed by Equation 10-10. These questions will be studied in detail in Chapter 14, where we will demonstrate that there is indeed a unique angular velocity vector for each motion, but that an instantaneous axis of rotation exists only under certain circumstances. For any two-dimensional motions except one of pure translation, an instantaneous center (instantaneous axis perpendicular to the plane of the motion) always exists.

Example

In the slider-crank mechanism shown in Figure 10-12, we would like to determine the relationship between the translational speed of the piston

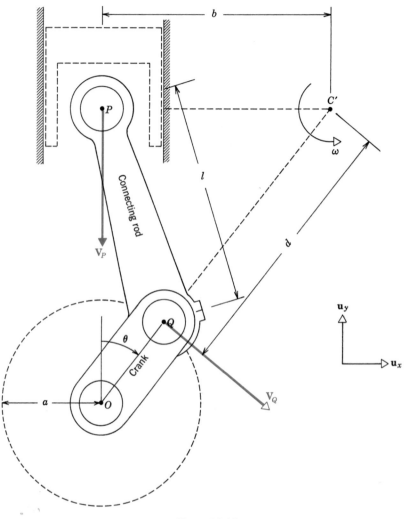

Figure 10-12

and the rotational speed of the crank. To this end, let us examine the connecting rod, and determine: (1) the location of its instantaneous center in terms of the crank position θ, (2) its angular velocity, and (3) the speed of the piston, in terms of the position and speed of the crank. By inspection, we can determine the directions of the velocities of the center of the wrist pin and the center of the crankpin. Since every point must move in a direction perpendicular to a line connecting the point and the instantaneous center, the center must be located at the point labeled C' in the figure. The angular velocity of the rod may be found in terms of the speed of the crankpin which, in turn, is given in terms of the angular speed of the crank.

$$\omega = \frac{v_Q}{d} = \frac{a\dot\theta}{d}$$

But the distance d may be found in terms of l, a, and θ:

$$d = \frac{\sqrt{l^2 - a^2 \sin^2 \theta}}{\cos \theta}$$

so that

$$\omega = \frac{a \cos \theta}{\sqrt{l^2 - a^2 \sin^2 \theta}}\, \dot\theta$$

The speed of the wrist pin may then be written in terms of the angular velocity just determined.

$$v_P = b\omega$$

Another geometry analysis gives us the distance b in terms of l, a, and θ.

$$b = a \sin \theta + \sqrt{l^2 - a^2 \sin^2 \theta} \ \tan \theta$$

The speed of the piston may now be determined in terms of l, a, θ, and $\dot\theta$ by combining the above results.

$$v_P = (a \sin \theta + \sqrt{l^2 - a^2 \sin^2 \theta} \ \tan \theta) \left(\frac{a \cos \theta}{\sqrt{l^2 - a^2 \sin^2 \theta}}\, \dot\theta \right)$$

$$= a \sin \theta \left(1 + \frac{\cos \theta}{\sqrt{l^2/a^2 - \sin^2 \theta}} \right) \dot\theta$$

It would probably be wise to check this result for the values of θ for which v_P can be seen from inspection.

Accelerations of Points on a Rigid Body.   Determination of acceleration is often facilitated through analysis of the angular velocity of a rigid body. For this purpose, we differentiate the velocity relationship (10-11), with the result

$$\dot{\mathbf{v}}_P = \dot{\mathbf{v}}_Q + \dot{\boldsymbol\omega} \times \mathbf{r}_{P/Q} + \boldsymbol\omega \times \dot{\mathbf{r}}_{P/Q}$$

As pointed out earlier, the velocity difference $\dot{\mathbf{r}}_{P/Q} = \mathbf{v}_{P/Q}$ may be expressed as $\boldsymbol\omega \times \mathbf{r}_{P/Q}$. With this and the definition $\mathbf{a} = \dot{\mathbf{v}}$, the above acceleration relationship may be written as

$$\boxed{\mathbf{a}_P = \mathbf{a}_Q + \dot{\boldsymbol\omega} \times \mathbf{r}_{P/Q} + \boldsymbol\omega \times (\boldsymbol\omega \times \mathbf{r}_{P/Q})} \qquad (10\text{-}12)$$

The vector $\dot{\boldsymbol{\omega}}$ is called the *angular acceleration* of the rigid body. In general, this vector may have any direction. However, in two-dimensional motion, in which the vector $\boldsymbol{\omega}$ remains perpendicular to the plane of the motion, $\dot{\boldsymbol{\omega}}$ will have only a "stretching" component. An example would be a turbine rotor mounted in fixed bearings, as its speed increases or decreases. On the other hand, a rotor inside a turbojet aircraft engine would have an angular velocity that would undergo a change in *direction* as the aircraft maneuvers; thus $\dot{\boldsymbol{\omega}}$ would have a "swinging" component in this case.

Example

Determine the acceleration of a point P attached to the rim of a 7-in. diameter rotor that is mounted in fixed bearings. The rotor speed increases according to

$$n = 1760 \text{ rpm } [1 - e^{-(0.8 \text{ s}^{-1})t}]$$

The angular velocity is given by

$$\boldsymbol{\omega} = \omega \, \mathbf{u}_z$$

where $\mathbf{u}_z$ is a unit vector parallel to the fixed axis of rotation, and the magnitude is given by

$$\omega = (1760 \text{ rpm})(2\pi \text{ rad/rev}) \left(\frac{1 \text{ min}}{60 \text{ s}} \right) [1 - e^{-(0.8 \text{ s}^{-1})t}]$$

$$= 184.3 \text{ rad/s } [1 - e^{-(0.8 \text{ s}^{-1})t}]$$

The angular acceleration is then

$$\dot{\boldsymbol{\omega}} = (0.8 \text{ s}^{-1})(184.3 \text{ rad/s}) \, e^{-(0.8 \text{ s}^{-1})t} \, \mathbf{u}_z$$
$$= (147.4 \text{ rad/s}^2) \, e^{-(0.8 \text{ s}^{-1})t} \, \mathbf{u}_z$$

In order to apply Equation 10-12, let us select point Q on the axis of rotation. Then, because this point does not move,

$$\mathbf{a}_Q = \mathbf{v}_Q = 0$$

The velocity of P will be

$$\boldsymbol{\omega} \times \mathbf{r}_{P/Q} = \omega(3.5 \text{ in.}) \, \mathbf{u}_\phi$$

where $\mathbf{u}_\phi$ is a unit vector in the tangential direction. Substitution into Equation 10-12 now yields:

$$\mathbf{a}_P = 0 + [(147.4 \text{ rad/s}^2)^{-(0.8 \text{ s}^{-1})t}] \, \mathbf{u}_z \times \mathbf{r}_{P/Q}$$
$$+ \omega \, \mathbf{u}_z \times [(3.5 \text{ in.})\omega \, \mathbf{u}_\phi]$$
$$= [(147.4 \text{ rad/s}^2)^{-(0.8 \text{ s}^{-1})t}](3.5 \text{ in})\mathbf{u}_\phi - (3.5 \text{ in})\omega^2 \, \mathbf{u}_\rho$$
$$= (516. \text{ in/s}^2)e^{-(0.8 \text{ s}^{-1})t} \, \mathbf{u}_\phi$$
$$- (1.19 \times 10^5 \text{ in/s}^2)[1 - ^{-(0.8 \text{ s}^{-1})t}]^2 \, \mathbf{u}_\rho$$

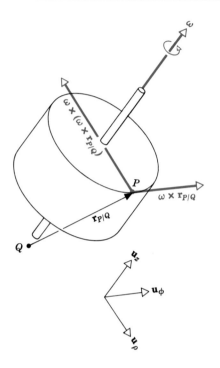

As is common in rotating machinery, the dominant component of acceleration is in the radial direction.

Reworking this example with reference to Equation 10-3b will enable you to unify some of the ideas that have appeared in different mathematical forms.

Example

The rotation rate $\dot{\theta}$ in the slider-crank mechanism of Figure 10-12 is constant. Evaluate the angular acceleration of the connecting rod and the acceleration of the piston.

The angular velocity of the connecting rod remains perpendicular to the plane of the figure, so that its derivative has the same orientation. Therefore the angular acceleration $\dot{\boldsymbol{\omega}}$ may be determined directly from the expression for angular velocity obtained earlier.

$$\dot{\boldsymbol{\omega}} = \dot{\omega}\, \mathbf{u}_z$$

$$= \frac{d}{dt}\left(\frac{a\cos\theta}{\sqrt{l^2 - a^2\sin^2\theta}}\,\dot{\theta}\right)\mathbf{u}_z$$

$$= -\frac{\left(\dfrac{l^2}{a^2} - 1\right)\sin\theta}{\left(\dfrac{l^2}{a^2} - \sin^2\theta\right)^{3/2}}\,\dot{\theta}^2\,\mathbf{u}_z \tag{a}$$

As point Q travels at constant speed in the circular path of radius a, its acceleration has the magnitude $a\dot{\theta}^2$ and is directed toward the point O. In terms of the horizontal and vertical unit vectors, this acceleration may be expressed as

$$\mathbf{a}_Q = -a\dot{\theta}^2(\sin\theta\ \mathbf{u}_x + \cos\theta\ \mathbf{u}_y) \tag{b}$$

The position vector from Q to P may be expressed similarly as

$$\mathbf{r}_{P/Q} = -a\sin\theta\ \mathbf{u}_x + \sqrt{l^2 - a^2\sin^2\theta}\ \mathbf{u}_y \tag{c}$$

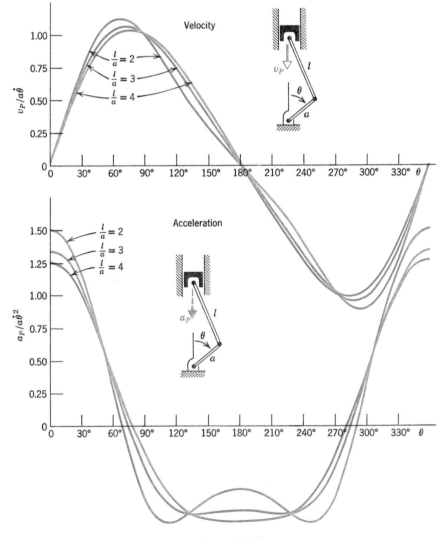

Figure 10-13

Now, the vector $\boldsymbol{\omega} \times \mathbf{r}_{P/Q}$ lies in the plane of the figure and is directed perpendicular to the line OP. Its magnitude is equal to $\omega\, l \sin (\pi/2)$. Then the vector $\boldsymbol{\omega} \times (\boldsymbol{\omega} \times \mathbf{r}_{P/Q})$ is directed opposite to $\mathbf{r}_{P/Q}$ and its magnitude is equal to $\omega[\omega l \sin \pi/2]\sin (\pi/2)$. Therefore we can express this term as

$$\boldsymbol{\omega} \times (\boldsymbol{\omega} \times \mathbf{r}) = -\omega^2\, \mathbf{r}_{P/Q} \qquad \text{(d)}$$

Substitution of the values given by Equations a, b, c, and d into Equation 10-12 leads to:

$$\mathbf{a}_P = -a\dot{\theta}^2 \left[\cos \theta + \frac{\dfrac{l^2}{a^2} (\cos^2 \theta - \sin^2 \theta) + \sin^4 \theta}{\left(\dfrac{l^2}{a^2} - \sin^2 \theta\right)^{3/2}} \right] \mathbf{u}_y$$

The variation of this acceleration with crank angle is shown in Figure 10-13b. It is interesting to compare the curves of Figure 10-13 with those from the Scotch yoke mechanism, shown in Figure 10-3.

You may find it instructive to evaluate the acceleration in this last example by direct differentiation of the velocity relationship on p. 43. Why would this procedure be valid in this case?

Problems

10-42   For each of the following rigid bodies moving in the plane, locate the instantaneous center and evaluate the quantities required.

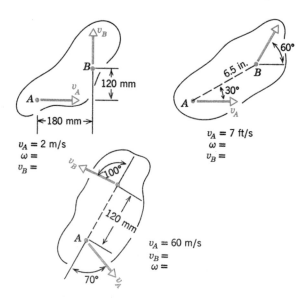

$v_A = 2$ m/s
$\omega =$
$v_B =$

$v_A = 7$ ft/s
$\omega =$
$v_B =$

$v_A = 60$ m/s
$v_B =$
$\omega =$

10-43   The rear wheels of a car spin on ice as the car moves directly forward. The angular velocity of the rear wheels is 450 rpm in the direction tending to make the car go forward. This point A of the tire in contact with the ice has a linear velocity of 15 mph to the left. Determine the angular velocity of the front wheels.

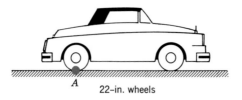

A
22-in. wheels

10-44   A bullet is fired through a rifled barrel 2.0 ft long and 0.30 in. in diameter. If the muzzle velocity is 2700 ft/s as the bullet leaves the barrel and the sprial lands of the barrel give it 2.4 complete turns as it passes through, determine the angular velocity of the bullet and the resultant acceleration of particles on the surface of the bullet as it leaves the barrel.

10-45   Give three different descriptions of the motion of the connecting rod in Figure 10-12 in terms of superpositions of translations and rotations about different points.

10-46   Make a graphical construction of the mechanism in Figure 10-12, for the case $l = 2.5a$, $\theta = 30°$. Determine the ratio v_P/v_Q graphically. For the value of ω determined by your scale for v_P and a, verify graphically that $v_{P/Q} = \omega \, r_{P/Q}$.

10-47   Use Equation 10-12 to derive Equation 10-3.

10-48   The rod AB moves in such a way that it remains in contact with the floor at B and with the corner at C. End B moves to the right with a speed v_0. Determine the angular velocity of the rod.

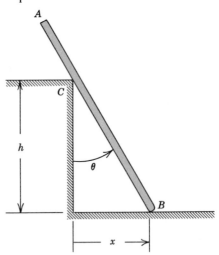

10-49 A wheel rolls along a surface without slipping.
   (a) Where is the instantaneous center?
   (b) What will be the path traced out in space by a succession of
       such points? (This path is called the *space centrode.*)
   (c) What will be the path traced out on the wheel, by a succession
       of points that become, successively, the instantaneous center?
       (This path is called the *body centrode.*)
   (d) Sketch the body centrode and space centrode for the con-
       necting rod of Figure 10-12.

10-50 The flywheel rotates with an approximately constant angular
   velocity of 2000 rpm. Estimate the velocity and acceleration of the
   piston at the instant shown.

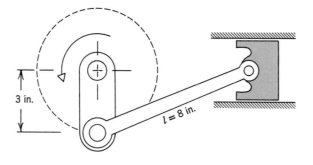

10-51 The crank rotates at an
   approximately   constant
   3600 rpm. Estimate the
   acceleration of the piston
   when the assembly is in
   the position shown.

10-52 A wheel rolls without slip
   along a smooth surface, its
   center moving with speed
   $v(t)$. Evaluate the acceleration of the point C' attached to the wheel.

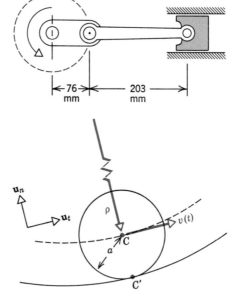

10-53 Verify the expression on p. 42 for the speed of the piston in Figure 10-12 for values of θ for which v_P can be evaluated by inspection.

10-54 Determine the speed of the piston of Figure 10-12 by an approach slightly different from that used in the example: Sketch the vectors $\mathbf{v}_{Q/O}$, $\mathbf{v}_{P/Q}$, and $\mathbf{v}_{P/O}$, showing their orientations relative to positions of the parts of the mechanism and their interrelationship $\mathbf{v}_{P/O} = \mathbf{v}_{P/Q} + \mathbf{v}_{Q/O}$. From this deduce $v_{P/O}$.

10-55 Still another way of determining the velocity of the piston in Figure 10-12 is to express its position as a function of θ and differentiate. Carry this out.

10-56 The device shown rotates about an OA at 250 rpm. The attached arm is designed to strike button B with each revolution. With what velocity (magnitude and direction) will it strike button B?

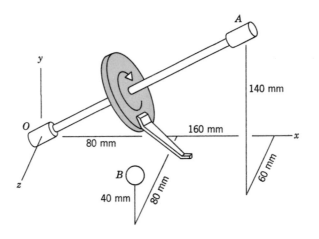

10-57 For the flyball mechanism shown in Figure 12-4, evaluate the velocity of a flyball and of the slider in terms of ψ, χ, $\dot{\psi}$, $\dot{\chi}$, L, and l.

10-58 The planet carrier arm drives the planet gear around the fixed sun gear, which is stationary.

(a) In terms of $\theta(t)$ and the two pitch circle radii a and b, what is the velocity of the center of the planet gear?

(b) Where is the instantaneous center for the planet gear?

(c) What is the angular velocity of the planet gear?

(d) Evaluate the velocity of point Q in terms of θ, $\dot{\theta}$, a, b, and ϕ.

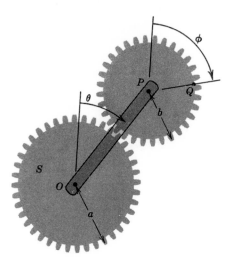

10-59 A disk with a 2-ft radius that rolls without slipping inside of a fixed circle with a 6-ft radius. The arm A has a constant clockwise angular velocity of 30 rpm. Determine the acceleration of point P when the disk is in the position shown.

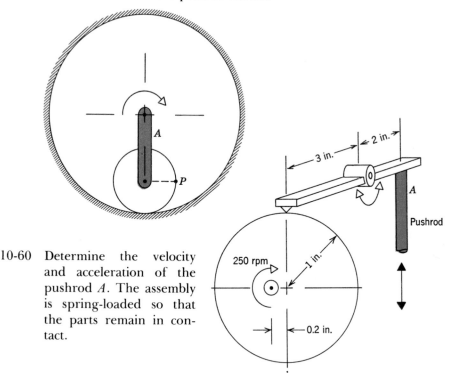

10-60 Determine the velocity and acceleration of the pushrod A. The assembly is spring-loaded so that the parts remain in contact.

10-61   The rotor of the Wankel rotary engine moves inside the housing
with its three apexes always in contact. On the side of the rotor
facing the viewer of the diagram is an internal gear that meshes
with the smaller gear that is fixed to the housing. On the opposite
side, depicted by dashed lines, is the drive shaft, having an offset
circular lobe that forms a bearing as it engages the rotor.

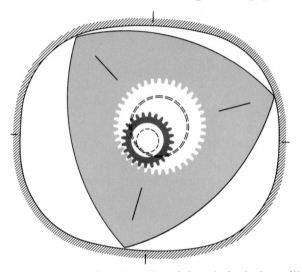

(a) What must be the ratio of the pitch circle radii of the gears so
that the rotor apexes move properly inside the housing?

(b) Locate the instantaneous center of the rotor.

(c) Determine the velocity of the center of the rotor and the
center of the offset lobe, in terms of the angular velocities of
the rotor and of the drive shaft and of the pitch circle radii.

(d) What is the ratio of the drive shaft angular velocity to the
rotor angular velocity?

10-62   Reference frame β is rotating relative to reference frame α with
angular velocity $\boldsymbol{\Omega}$. A vector $\mathbf{A}$ is constant in reference frame β,
so that $\overset{\beta}{\dot{\mathbf{A}}} = \mathbf{0}$. Show that $\overset{\alpha}{\dot{\mathbf{A}}} = \boldsymbol{\Omega} \times \mathbf{A}$.

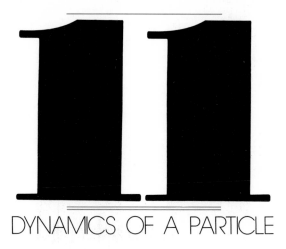

DYNAMICS OF A PARTICLE

We now turn to the *kinetics* of a moving particle. This part of the study of mechanics is concerned with the implications of Newton's second law of motion, as it relates the motion of a particle to the associated forces.

11–1

WRITING NEWTON'S SECOND LAW. To obtain a proper expression of Newton's second law of motion for a particular problem, we must apply the procedures described in Chapters 2, 3, and 4 in handling the forces, and the procedures described in Chapters 9 and 10 in handling the acceleration of the particle.

Example

The pendulum shown in Figure 11-1a consists of a particle of mass m, suspended from a fixed point by a rod of negligible mass. Because of friction, the hinge transmits a couple of moment M. In terms of this moment, the angle $\phi(t)$, and the physical constants l, m, and g, write expressions of Newton's second law.

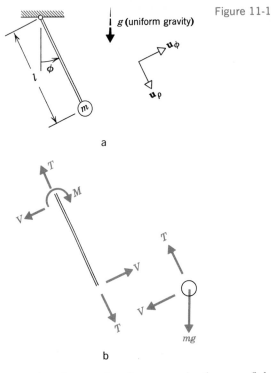

Figure 11-1

a

b

As always, the first step is the careful construction of appropriate free-body diagrams, as shown in Figure 11-1*b*. Since its mass is negligible, the sum of forces and moments acting on the rod must be zero, even though its velocities vary. Force equilibrium of the rod has already been accounted for in the labeling of the free body diagram; moment equilibrium requires that

$$V = \frac{M}{l} \tag{a}$$

From the free-body diagram of the particle, we see that the force acting on it may be expressed as

$$\mathbf{f} = (mg \cos \phi - T) \, \mathbf{u}_\rho + (-mg \sin \phi - V) \, \mathbf{u}_\phi$$

Referring to Equation 10-7, we can express the acceleration of the particle as

$$\mathbf{a} = -l\dot{\phi}^2 \, \mathbf{u}_\rho + l\ddot{\phi} \, \mathbf{u}_\phi$$

Newton's second law can now be written as

$$(mg \cos \phi - T)\mathbf{u}_\rho + (-mg \sin \phi - V)\mathbf{u}_\phi = m[-l\dot{\phi}^2 \, \mathbf{u}_\rho + l\ddot{\phi} \, \mathbf{u}_\phi]$$

The equivalent component equations are

$$mg \cos \phi - T = -ml\dot{\phi}^2 \tag{b}$$
$$-mg \sin \phi - V = ml\ddot{\phi} \tag{c}$$

In Equations a, b, and c, we now have a mathematical expression of the laws of force and motion interaction for this system. In subsequent sections we will examine procedures for predicting the motion and forces from these equations.

Problems

11-1 Two blocks of mass $m_1 = 4$ kg and $m_2 = 3$ kg are sliding down the inclined plane shown. The coefficient of kinetic friction between block m_2 and the plane is $\mu = 0.1$. Determine the acceleration of the blocks and the force in the rod connecting them when the coefficient of friction between block m_1 and the plane is
(a) $\mu = 0.05$.
(b) $\mu = 0.10$.
(c) $\mu = 0.20$.

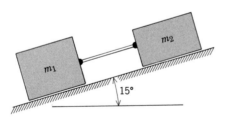

11-2 Same as Problem 11-1 except that the blocks are interchanged.
11-3 A 200-lb man is riding in an elevator. Determine the force that the man exerts on the floor when the elevator is:
(a) Accelerating upward at a rate of 3 ft/s².
(b) Accelerating downwards at a rate of 3 ft/s².
11-4 A 2.2 Mg elevator is supported by a cable. Determine the tension in the cable when the elevator is:
(a) Accelerating upwards at a rate of 1 m/s².
(b) Accelerating downwards at a rate of 1 m/s².

11-5 A body of mass $m_1 = 3$ slugs and a body of mass $m_2 = 4$ slugs are connected by a cable as shown. Determine the acceleration of both bodies and the tension in the cable. Friction is negligible.

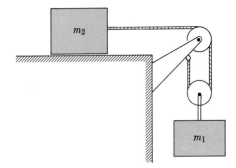

11-6 Same as Problem 11-5 except that the coefficient of kinetic friction is $\mu = 0.2$.

11-7 A block of mass $m_1 = 3$ slugs rests on a block of mass $m_2 = 4$ slugs as shown. Determine the maximum force P that may be applied to the lower block before the upper block begins to slide off. The coefficient of static friction between the blocks is $\mu = 0.2$. Friction between the table and m_2 is negligible.

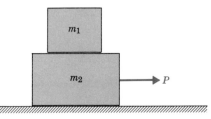

11-8 Same as Problem 11-7 except that the coefficient of kinetic friction between m_2 and the table is $\mu = 0.1$.

11-9 An automobile of mass m starts from rest at time $t = 0$ and moves in a straight line as torque T is delivered to the rear wheels through the rear axle. The torque varies according to $T = T_0\, e^{-kt}$ and the diameter of the rear wheels is D. Determine the acceleration, velocity, and displacement of the car at time t.

11-10 A pendulum hangs inside a truck that is accelerating horizontally at the rate of 2.45 m/s². Determine the equilibrium position of the pendulum in the truck.

11-11 A 190-lb man is being hoisted into a helicopter, which is climbing in a straight line at 45° from the vertical and accelerating at

10 ft/s². The cable is being reeled in at 4 ft/s, uniformly. There is no pendulumlike swinging of the man. Determine the tension in the cable and its inclination from the vertical.

11-12 A tank containing water slides down an inclined plane under the action of gravity. The angle between the plane and the horizontal is α. Find the angle between the water surface and the horizontal (a) where the surface is frictionless; (b) where the coefficient of friction between tank and surface is μ.

11-13 An automobile manufacturer claims for his car an acceleration from 15 to 50 mph in 13 s in high gear. If a glass U-tube with vertical legs 3 ft apart is partly filled with water and used as an accelerometer, what is the difference in level of the two legs for this constant acceleration? For an initial speed of 15 mph, how fast would the car be going at the end of 13 s if the difference in level were $\frac{1}{2}$ in. larger? If it were $\frac{1}{2}$ in. smaller?

11-14 A boy is twirling an object of mass m on a string of length ρ in a circular path in a vertical plane. Assuming the boy's hand is stationary, determine the required value of v such that the string does not become slack when the object is at the top of the circle.

11-15 An automobile is traveling around a curve. The radius of curvature of the path of the car is ρ. The magnitude of the velocity **v** is constant. Determine the bank angle such that the friction force at the roadway is zero.

11-16 An earth satellite moves in a circular orbit above the earth. The period for one complete revolution is 90 min. (a) Determine the acceleration of gravity at the satellite, in terms of the altitude and period. (b) Determine the altitude of the satellite.

11-17 Determine the velocity required for a circular orbit 100 mi above the surface of the moon. The radius of the moon is 1080 mi and the acceleration at the surface of the moon caused by its gravity is 5.31 ft/s^2.

11-18 Estimate the altitude and velocity a satellite must have to remain in a stable circular orbit around the earth such that it is always approximately over the same meridian of longitude.

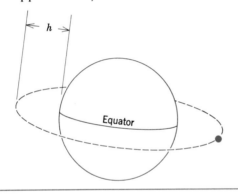

11-2

A CLASSIFICATION OF PROBLEMS. The known quantities and those quantities that a problem requires our predicting may appear on either side of the equation $\mathbf{f} = m\mathbf{a}$. A classification of problems, based on whether the motion is completely given, is helpful in recognizing what kinds of steps will be required to solve the problem.

Motion-Given Problems. If the motion of a system is given and the unknown quantities to be predicted are force components, an analysis of accelerations and construction of appropriate free-body diagrams reduce the problem to one of straightforward differentiation and algebra.

Example

Suppose that the hinge of the pendulum in Figure 11-1 has negligible friction, and we know from measurements that the motion of the pendulum is given by

$$\phi = \phi_0 \cos \omega_0 t \qquad \text{(d)}$$

where the amplitude ϕ_0 and angular frequency ω_0 are known quantities. Determine the tension in the rod.

The force and acceleration analysis carried out previously produced Equation b, in which the desired tension is now the only unknown. This yields

$$T = mg \cos \phi + ml\dot{\phi}^2$$
$$= mg \cos (\phi_0 \cos \omega_0 t) + ml(-\omega_0\phi_0 \sin \omega_0 t)^2$$

Force-Given Problems. If the forces are known, either explicitly or in terms of the configuration and/or motion of the system, the expression of Newton's laws of motion may be expected to result in differential equations that must be integrated. Such problems range from fairly simple to very formidable.

Example

Suppose that the friction moment in the hinge of the above example is proportional to the square of the rotation rate,

$$M = \alpha \dot{\phi} |\dot{\phi}| \qquad \text{(e)}$$

and we are to predict the motion of the pendulum.

Substitution of Equations a and e into Equation c results in the differential equation:

$$\ddot{\phi} + \frac{\alpha}{ml^2} \dot{\phi} |\dot{\phi}| + \frac{g}{l} \sin \phi = 0 \qquad \text{(f)}$$

Integration of this differential equation would give the motion we seek; however, no solution to this equation is available in terms of known functions.

In the case in which the friction is negligible ($\alpha = 0$) and the angle ϕ is restricted to the range in which the approximation

$$\sin \phi = \phi - \frac{\phi^3}{3!} + \cdot \cdot \cdot \approx \phi$$

is valid, the equation reduces to

$$\ddot{\phi} + \frac{g}{l} \phi = 0$$

You may verify by substitution that Equation d is a solution, with $\omega_0 = \sqrt{g/l}$.

In the following sections we will examine some approaches to integrating equations that commonly arise in dynamics. Procedures for somewhat more complex dynamics systems are presented in Chapter 18.

Problems

Before solving, classify each of the problems in this section as motion-given or force-given.

11-19 Show that for negligible friction and small ϕ, the tension in the rod in Figure 11-1 is given by

$$T = mg\left[1 + \frac{\phi_0^2}{4}(1 - 3\cos 2\omega_0 t)\right]$$

11-20 Derive the differential equations of motion for the spherical pendulum. Suggestion: Equation 10-8c may be helpful.

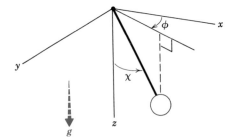

11-21 The spherical pendulum of the previous problem is observed to rotate around the vertical with χ constant. Determine the rate of rotation $\dot{\phi}$ and the tension in the rod.

11-22 The thrust of a boat's propeller is constant, T_0. The drag from the surrounding water is proportional to the speed of the boat, the constant of proportionality being c. Determine the equation that these assumptions imply, governing the boat's speed, v.

11-23 A particle of mass m moves along a helical guide under the influence of gravity. The axis of the helix is vertical, and there is negligible friction between the particle and the guide. Show that its speed will vary as postulated in the example of pp. 24–25. Evaluate b in terms of g and the helix geometry. Evaluate the reaction between the guide and the particle.

11-24 The system as described in Problem 11-23 has Coulomb friction between the particle and the guide. In terms of the coefficient of friction μ, m, g, c, and α, determine the differential equation governing $v(t)$, that these assumptions imply.

11-25   The man is being reeled into the rescue helicopter shown in Figure 10-9 at a constant rate of 0.8 m/s. The acceleration $\mathbf{a}_B$ shown is constant. Write the differential equation that governs $\phi(t)$.

11-26   The force exerted by the spring is proportional to its displacement, $f_s = k(\text{displacement})$. Write the differential equation for $x(t)$, where the origin for x is:
(a)   At the position where the spring is unstressed.
(b)   At the position where the spring force balances the gravitational force.

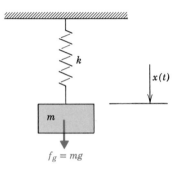

11-27   The cart has a prescribed motion given by the displacement $s(t)$. The spring exerts a force proportional to its extension, $f_s = kx$. Write the equation of motion that governs the relative displacement $x(t)$ of the block.

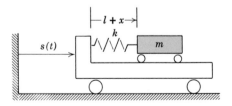

11-28   A 45-Mg aircraft is in a steady climbing turn at 180 m/s, such that its rate of climb is 1800 m/min and the radius of its helical path is 2400 m. The drag force (acting parallel to the flight path) is

50 kN. Determine the magnitude of lift (perpendicular to the flight path) and thrust (parallel to the flight path).

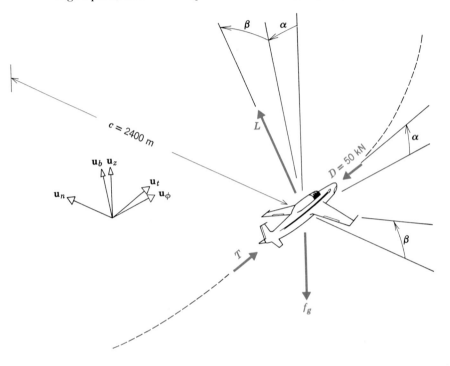

11-29   A 200-g valve is about to be lifted by a bump on a cam moving at 15 m/s. The bump extends over a 40 mm length and has a shape given by $y = 6$ mm$[1 - \cos(\pi x/20$ mm$)]$. The stiffness of the spring is $k = 180$ kN/m. How much preload or initial compression must the spring have to prevent loss of contact with the cam?

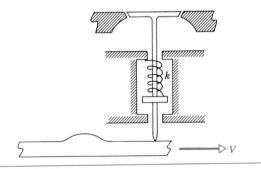

11-3

SOLVABLE ONE-DIMENSIONAL PROBLEMS. If the force causing
rectilinear motion of a particle obeys one of the three special forms listed
in the following, the integration can be carried out as described.

Force an Explicit Function of Time. When the force acting on the particle is a given function of time, $F(t)$, the velocity and position can be determined by two successive integrations of Newton's second law:

$$m\dot{v} = F(t) \tag{11-1a}$$

$$v = v_0 + \frac{1}{m} \int F(t)\, dt \tag{11-1b}$$

$$x = \int v\, dt = x_0 + v_0 t + \frac{1}{m} \int \int F(t)\, dt\, dt \tag{11-1c}$$

The constants of integration v_0 and x_0 are normally determined from
knowledge of the initial values of velocity and displacement. An example
is the vertical motion of a particle under the influence of the gravitational
attraction near the surface of the earth. Taking z as the height of the particle above the surface of the earth, Newton's second law may be expressed as

$$m\ddot{z} = -mg$$

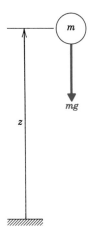

A simple integration gives the velocity:

$$v = v_0 - gt$$

and another gives the height:

$$z = z_0 + v_0 t - \tfrac{1}{2}gt^2$$

The constants of integration v_0 and z_0 are the initial speed and height, respectively.

Force a Function only of Velocity. When the force acting on the particle is a function of the velocity, v, the variables v and t can be separated in Newton's second law, and integration then be carried out. First, the equation

$$m \frac{dv}{dt} = F(v)$$ (11-2a)

is rearranged as

$$\frac{m \, dv}{F(v)} = dt$$

Integration then gives

$$\int \frac{m \, dv}{F(v)} = t_0 + t$$ (11-2b)

where t_0 is a constant of integration. We now have t as a function of v; the inverse relationship will give velocity as a function of time. The distance x can be determined by a further integration.

$$x = x_0 + \int v \, dt$$

Substitution of $m \, dv/F$ for dt (above) gives a means of computing distance directly from the force-velocity relationship $F(v)$.

$$x = x_0 + \int \frac{mv \, dv}{F(v)}$$ (11-2c)

As an example, consider a particle under the influence of gravitational attraction and a drag force proportional to velocity. With the selection of coordinate as shown in the diagram, Newton's second law is expressed as

$$m\ddot{z} = m \frac{dv}{dt} = -mg - cv$$

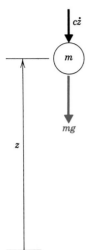

Or, in the rearranged form,

$$\frac{dv}{g + \dfrac{c}{m} v} = -dt$$

Integration yields

$$\frac{m}{c} \log \left(g + \frac{cv}{m} \right) = t_0 - t$$

and inversion of this relationship gives the velocity.

$$v(t) = \frac{m}{c} \left[-g + e^{c(t_0 - t)/m} \right]$$

If the initial velocity is given, $v(0) = v_0$, the constant of integration t_0 may be evaluated by substitution into the above.

$$v_0 = \frac{m}{c} \left[-g + e^{ct_0/m} \right]$$

Solving this for the constant $e^{ct_0/m}$ and substituting above yield

$$v(t) = -\frac{mg}{c} \left[1 - \left(1 + \frac{cv_0}{mg} \right) e^{-ct/m} \right]$$

Note that as ct/m becomes large, the velocity approaches the constant "terminal" velocity, $-mg/c$, where drag and gravitational forces balance one another. Another integration of this last equation will yield an expression for the height, $z(t) = \int v \, dt$.

Force a Function only of Distance. When the force is a function only of distance x, integration can be carried out as follows. Multiplication by velocity of each member of Newton's second law,

$$\boxed{m \frac{dv}{dt} = F(x)}$$

(11-3a)

results in

$$mv \frac{dv}{dt} = F(x) \frac{dx}{dt}$$

Now, if the given force-distance law is integrated with respect to x and the result is called $-V(x)$:

$$\int F(x)\, dx = -V(x)$$

the above may be rewritten as

$$\frac{d}{dt} \left[\frac{mv^2}{2} + V(x) \right] = 0$$

and integrated

$$\boxed{\frac{mv^2}{2} + V(x) = E}$$

(11-3b)

The constant of integration E may be determined from the initial values of velocity and displacement. This done, the velocity of the particle is then determined from the above relationship.

$$v = \frac{dx}{dt} = \pm \sqrt{\frac{2}{m} [E - V(x)]}$$

Finally, the distance x may be determined after separating the variables in the last equation.

$$\frac{dx}{\pm \sqrt{\frac{2}{m} [E - V(x)]}} = dt$$

$$\boxed{\pm \int \frac{dx}{\sqrt{\frac{2}{m} [E - V(x)]}} = t_0 + t}$$

(11-3c)

For illustration, consider the motion of a particle, under the action of a spring that produces a restoring force proportional to displacement.

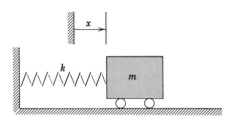

The proportionality constant is k (SI units: N/m). The displacement x is measured from the position of static equilibrium, so that the magnitude of the restoring force is kx. Newton's second law for this particle is then

$$m\dot{v} = F(x) = -kx$$

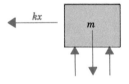

Integration of the force-displacement relationship yields

$$V(x) = -\int F(x)\ dx = \tfrac{1}{2}kx^2$$

Thus Equation 11-3b becomes, for this case,

$$\frac{mv^2}{2} + \frac{kx^2}{2} = E$$

Solving for the velocity results in

$$v = \frac{dx}{dt} = \pm \sqrt{E - \frac{kx^2}{2}} \sqrt{\frac{2}{m}}$$

from which the variables may be separated:

$$\pm \frac{dx}{\sqrt{\frac{2E}{k} - x^2}} = \sqrt{\frac{k}{m}}\ dt$$

Integration then gives

$$\cos^{-1}\left(\frac{x}{\sqrt{2E/k}}\right) = \sqrt{\frac{k}{m}}\ t + \theta$$

in which θ is a constant of integration. This may be inverted to yield the displacement in terms of time as

$$x = \sqrt{\frac{2E}{k}}\ \cos(\omega_0 t + \theta)$$

in which

$$\omega_0 = \sqrt{\frac{k}{m}}$$

represents the angular frequency of oscillation. Observe that the frequency in this case depends only on the mass of the block and the stiffness of the spring, and not on the manner in which the motion is initiated. On the other hand, the amplitude of motion $\sqrt{2E/k}$ and the phase θ depend on the constants of integration, which in turn depend on the initial values of displacement and velocity.

It is instructive to compare this result with the analysis of the Scotch yoke mechanism of Figure 10-3.

Problems

11-30   A falling body is observed to fall 60 ft in 1 s. How far did it fall in the preceding second? How far will it fall in the next second?

11-31   A body is dropped near the surface of the earth. Neglecting air resistance, determine how far it falls during each of the first 4 s. If the body were dropped near the surface of the moon ($g = 1.62$ m/s^2), how far would it fall during each of the first four seconds?

11-32   An object A is dropped at a height of 78.5 m above the ground. An object B is dropped 2 s later. Determine the distance d between the starting elevations of A and B such that A and B hit the ground at the same time. What are the elevation and velocity of A when B is released?

11-33   A baseball is hit at an angle of inclination of 30° and at an elevation of 4 ft. The batter desires to have the ball clear a fence that is 300 ft away and 14 ft high. Estimate the initial velocity that he must impart to the ball. Is your estimate high or low?

11-34   A ball rolls off a horizontal table of height 760 mm and strikes the floor 500 mm from the edge of the table. What was its initial speed?

11-35   A basketball player releases the ball at a height of 2.4 m above the floor. The height of the basket is 10 ft. With what initial velocity should the player release the ball in order to have it swish through the hoop from a distance of 20 ft when the player releases the ball at an initial angle of (a) 30°, (b) 45°, and (c) 60°?

11-36   A 600-ton train moving up a grade of 3 ft in 100 ft has an initial speed of 5 mph. Assuming the train rolling resistance to be 15 lb/ton and the engine to provide a constant drawbar pull of 60,000 lb, estimate the time required for the train to reach a speed of 25 mph.

11-37 A 54.5 Mg railroad car is moving at 1.25 m/s down a grade which drops 2 m in 100 m. Its rolling resistance is 0.07 N/kg. Estimate the time required to stop the car by means of a force P of 5000 lb, applied as shown.

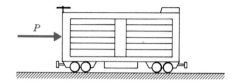

11-38 A certain towboat and tow load combination have a net thrust (i.e., total thrust minus total drag) versus velocity capability as shown. How much time and what distance will be required to accelerate the tug and tow from an initial speed v_0 to a higher speed v? Together they displace 80,000 lb.

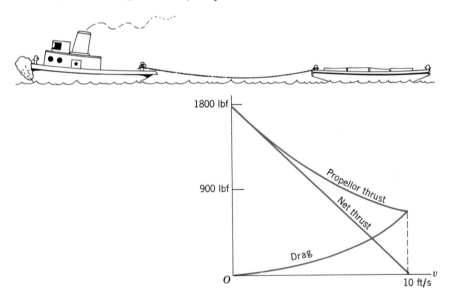

11-39 Use Equation 11-2c to compute the height of the particle that is launched vertically with initial velocity v_0 into a medium providing drag force proportional to velocity.

11-40 Sketch curves showing velocity and height for the previous problem.

11-41 A particle of mass m falls toward the earth (constant gravity) through a medium that provides a drag force proportional to the square of the velocity; the constant of proportionality is α N·s²/m². Determine the velocity from the time the particle is dropped from rest.

11-42   The particle of the previous problem is launched vertically upward with an initial velocity v_0. Determine its velocity and height as a function of time, and sketch curves showing how these vary with time.

11-43   The constant gravitational force acting on a particle was considered to be a special case of a force dependent on time, $F(t)$ (pp. 63–64). We could also consider $-mg$ to be a special case of $F(v)$ or of $F(x)$. Use the procedures for these cases (pp. 64–66) to integrate the equation of motion $m\ddot{z} = -mg$.

11-44   The spring in the example on p. 67 has a stiffness of 100 N/m. The body to which it is attached has a mass of 2 kg. What will be the period of oscillation?

11-45   The spring-mass system in the example on p. 67 is set in motion with the initial conditions $x(0) = 0$, $\dot{x}(0) = v_0$. Determine the expression for $x(t)$.

11–4

NUMERICAL INTEGRATION.   Problems that do not lend themselves to solution in terms of known functions can usually be handled by numerical procedures such as that described in this section. Many refinements to this approach have been developed and made practicable by modern digital computers. However, numerical integration has shortcomings that recommend its use only where analytical methods fail.

The procedure results in a numerical value for one or more dynamic variables at each of a set of discrete values of the independent variable t. With a sufficient number of such values, curves may be plotted and examined, and some understanding of the dynamic behavior of the system may be achieved.

A Predictor-Corrector Procedure.   Consider the differential equation

$$\frac{d^2x}{d\tau^2} = f\left(\tau, x, \frac{dx}{d\tau}\right) \tag{11-4}$$

The special cases in which the function f contains only τ, only x, or only $dx/d\tau$ can be analyzed by the procedures described in the preceding section. However, for a more general case, such as Equation f on page 59, an approach such as described here is often called for. For later convenience, let us rewrite the second-order equation (11-4) as an equivalent pair of first-order equations. This may be accomplished by defining the variable y as

$$\frac{dx}{d\tau} = y \tag{11-4a}$$

Then Equation 11-4 may be written as

$$\frac{dy}{d\tau} = f(\tau, x, y) \tag{11-4b}$$

Referring to Figure 11-2, we note that the slope of the $y - \tau$ curve at the

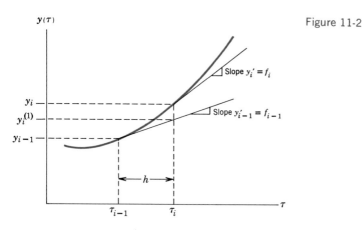

Figure 11-2

point τ_i is given by

$$y_i' = f(\tau_i, x_i, y_i) \equiv f_i$$

where the subscript i refers to a particular value of τ and corresponding values of x, y, and f. We see further from the figure that a first approximation to y_i can be written as

$$y_i^{(1)} = y_{i-1} + h f_{i-1} \tag{11-5a}$$

However, this approximation could be improved by using the average of the slopes f_{i-1} and f_i instead of f_{i-1} alone.

$$y_i = y_{i-1} + h \frac{f_{i-1} + f_i}{2}$$

$$= \left(y_{i-1} + \frac{h}{2} f_{i-1} \right) + \frac{h}{2} f_i$$

$$= Y_{i-1} + \frac{h}{2} f_i \tag{a}$$

A similar approximation can be written for determining changes in x in terms of values of y.

$$x_i = x_{i-1} + h \frac{y_{i-1} + y_i}{2}$$

$$= \left(x_{i-1} + \frac{h}{2} y_{i-1} \right) + \frac{h}{2} y_i$$

$$= X_{i-1} + \frac{h}{2} y_i \tag{b}$$

We will use (a) and (b) to determine, successively, values of $y_1, x_1, y_2, x_2, y_3,$ x_3, and so forth, starting with initial values y_0 and x_0.

Suppose now that we have computed values of y_{i-1} and x_{i-1}. Unfortunately, Equations a and b do not give us the values y_i and x_i directly, because only the terms $Y_{i-1} = [y_{i-1} + (h/2)f_{i-1}]$ and $X_{i-1} = [x_{i-1} + (h/2)y_{i-1}]$ in the right-hand sides are known at this stage of the computation. To circumvent this difficulty, let us use the first approximation (11-5a), and a similar first approximation for x_i,

$$x_i^{(1)} = x_{i-1} + h\, y_{i-1} \tag{11-5b}$$

to compute a corresponding first approximation to f_i:

$$f_i^{(1)} = f[\tau_i, x_i^{(1)}, y_i^{(1)}]$$

This may then be used in Equation a to compute a second approximation to y_i:

$$y_1^{(2)} = Y_{i-1} + \frac{h}{2}f_i^{(1)}$$

and this value in Equation b to compute a second approximation to x_i.

$$x_i^{(2)} = X_{i-1} + \frac{h}{2}y_i^{(2)}$$

We then repeat this procedure, as summarized in the following,

$$y_i^{(j)} = Y_{i-1} + \frac{h}{2}f_i^{(j-1)} \tag{11-6a}$$

$$x_i^{(j)} = X_{i-1} + \frac{h}{2}y_i^{(j)} \tag{11-6b}$$

$$f_i^{(j)} = f(\tau_i, x_i^{(j)}, y_i^{(j)}) \tag{11-6c}$$

until two successive approximations yield values that differ by no more than some selected tolerance. If more than two such iterations are necessary, it usually indicates that a smaller increment h should be used.

After the values y_i and x_i are determined, the procedure is repeated to evaluate y_{i+1} and x_{i+1}.

Equations 11-5 are known as *predictors,* inasmuch as they predict, in terms of values at τ_{i-1}, the values at τ_i. Equations 11-6 are known as *correctors,* which yield improved approximations for the values at τ_i in terms of the estimates given by the predictors or previous application of the correctors. The approach is thus called a *predictor-corrector technique.*

Example

Determine the motion of a quadratically damped pendulum (see p. 59) governed by the differential equation

$$\ddot{\phi} + \frac{\alpha}{ml^2}\,\dot{\phi}|\dot{\phi}| + \frac{g}{l}\sin\phi = 0$$

and the initial conditions

$$\phi(0) = 1.5; \qquad \dot{\phi}(0) = 0.5\ \text{s}^{-1}$$

For a reason to be pointed out in the next section, let us first introduce a dimensionless measure of time by the definition

$$\tau = \sqrt{\frac{g}{l}}\,t$$

Using the chain rule,

$$\frac{d\phi}{dt} = \frac{d\tau}{dt}\frac{d\phi}{d\tau} = \sqrt{\frac{g}{l}}\,\phi'$$

the governing equations may be written in terms of the new independent variable as:

$$\phi''(0) + \frac{\alpha}{ml^2}\,\phi'|\phi'| + \sin\phi = 0$$

$$\phi(0) = 1.5 \qquad \phi'(0) = (0.5\ \text{s}^{-1})\sqrt{\frac{l}{g}}$$

With $\phi = x$, the equivalent first-order equations are

$$\frac{dy}{d\tau} = f(x,y) = -\sin x - \left(\frac{\alpha}{ml^2}\right)y|y|$$

$$\frac{dx}{d\tau} = y$$

$$y(0) = (0.5\ \text{s}^{-1})\sqrt{\frac{l}{g}}\ ; \qquad x(0) = 1.5$$

The computation will be illustrated for the following values of the parameters:

$$\frac{\alpha}{ml^2} = 0.25, \qquad \sqrt{\frac{l}{g}} = 1.80\ \text{s}$$

and for equal intervals of τ, $h = 0.10$.

In the first interval, $0 < \tau \leqslant 0.1$, the computation proceeds as follows.

$$f_0 = -\sin x_0 - (0.25)y_0|y_0|$$
$$= -\sin 1.50 - 0.25(0.90)|(0.90)|$$
$$= -1.199\ 995$$

$$Y_0 = y_0 + \frac{h}{2} f_0$$

$$= 0.90 + \frac{(0.1)}{2} (-1.199\ 995)$$

$$= 0.840\ 000$$

$$X_0 = x_0 + \frac{h}{2} y_0$$

$$= 1.50 + \frac{(0.1)}{2} (0.9)$$

$$= 1.545\ 000$$

$$y_1^{(1)} = y_0 + hf_0$$
$$= (0.90) + (0.1)(-1.199\ 995)$$
$$= 0.780\ 001$$

$$x_1^{(1)} = x_0 + hy_0$$
$$= (1.50) + (0.1)(0.90)$$
$$= 1.590\ 000$$

$$f_1^{(1)} = -\sin x_1^{(1)} - (0.25)y_1^{(1)}\ |y_1^{(1)}|$$
$$= -\sin (1.590\ 000) - (0.25)(0.780\ 001)|0.780\ 001|$$
$$= -1.151\ 916$$

$$y_1^{(2)} = Y_0 + \frac{h}{2} f_1^{(1)}$$

$$= 0.840\ 000 + (0.05)(-1.151\ 916)$$
$$= 0.782\ 404$$

$$x_1^{(2)} = X_0 + \frac{h}{2} y_1^{(2)}$$

$$= 1.545\ 000 + (0.05)(0.782\ 404)$$
$$= 1.584\ 120$$

$$f_1^{(2)} = -\sin x_1^{(2)} - (0.25)y_1^{(2)}|y_1^{(2)}|$$
$$= -\sin (1.584\ 120) - (0.25)(0.782\ 404)|0.782\ 404|$$
$$= -1.152\ 950$$

The next iteration yields

$$y_1^{(3)} = 0.782\ 353$$
$$x_1^{(3)} = 1.584\ 118$$

which agrees satisfactorily with $y_2^{(2)}$ and $x_2^{(2)}$. We then proceed to the next interval, where

$$Y_1 = 0.724\ 705$$
$$X_1 = 1.623\ 235$$
$$y_2^{(1)} = 0.667\ 058$$
$$x_2^{(1)} = 1.662\ 353$$
$$f_2^{(1)} = -1.107\ 053$$
$$y_2^{(2)} = 0.669\ 353$$
$$x_2^{(2)} = 1.656\ 703$$

.

.

.

The results of the computations for the first 30 intervals are listed in Table 11-1. In no interval were more than two iterations required to meet the criterion

$$|x_i^{(j+1)} - x_i^{(j)}| < 0.0005\ |x_i^{(j)}|$$

The values from the table are plotted in Figure 11-3. From this curve we observe the oscillatory motion that one would expect intuitively.

Table 11-1

τ_i	$\sqrt{g/l}\ y_i$	x_i	τ_i	$\sqrt{g/l}\ y_i$	x_i
0	0.900 000	1.500 000	2.0	−1.008 288	1.276 383
0.2	0.782 353	1.584 118	2.1	−1.075 064	1.172 215
0.3	0.669 289	1.656 700	2.2	−1.134 277	1.061 748
0.3	0.560 488	1.718 189	2.3	−1.184 858	0.945 792
0.4	0.455 486	1.768 987	2.4	−1.225 816	0.825 258
0.5	0.353 724	1.809 448	2.5	−1.256 302	0.701 152
0.6	0.254 565	1.839 862	2.6	−1.275 669	0.574 554
0.7	0.157 327	1.860 457	2.7	−1.283 513	0.446 594
0.8	0.061 295	1.871 388	2.8	−1.279 712	0.318 433
0.9	−0.034 234	1.872 741	2.9	−1.264 428	0.191 226
1.0	−0.129 605	1.864 549	3.0	−1.238 095	0.066 100
1.1	−0.224 730	1.846 832	3.1	−1.201 402	−0.055 875
1.2	−0.319 392	1.819 626	3.2	−1.155 251	−0.173 707
1.3	−0.413 321	1.782 990	3.3	−1.100 668	−0.286 503
1.4	−0.506 173	1.737 016	3.4	−1.038 752	−0.393 474
1.5	−0.597 512	1.681 831	3.5	−0.970 625	−0.493 943
1.6	−0.686 792	1.617 616	3.6	−0.897 381	−0.587 344
1.7	−0.773 349	1.544 609	3.7	−0.820 036	−0.673 214
1.8	−0.856 399	1.463 122	3.8	−0.739 502	−0.751 191
1.9	−0.935 045	1.373 550	3.9	−0.656 567	−0.820 995
2.0	−1.008 288	1.276 383	4.0	−0.571 891	−0.882 418

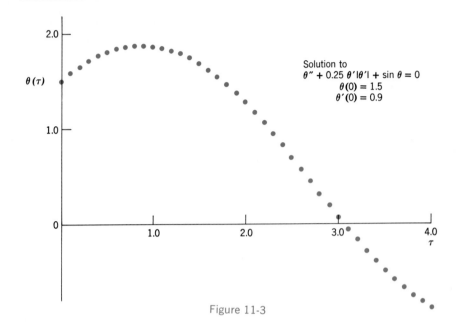

$\theta(\tau)$

Solution to
$\theta'' + 0.25\ \theta'|\theta'| + \sin\theta = 0$
$\theta(0) = 1.5$
$\theta'(0) = 0.9$

Figure 11-3

Numerical vis-à-vis Analytical Integration.   The chief advantage of methods such as described in the preceding is that they provide the only means through which the differential equations of many models may be integrated. The effort required to develop the computer programs is well compensated by the ease with which values can be written and automatically plotted. Inaccuracies inherent in the method are only occasionally a serious problem, and they are becoming less so as refinements in the methods are developed and the capacity of computing machines is improved. Moreover, computing costs have diminished to the point where they are generally negligible compared with the cost of the analyst's time to interpret the results of his computations.

It is the increased effort required to *interpret results* that dictates that we resort to numerical integration only where analytical methods fail. Note, for example, that the results in Figure 11-3 are for only *one* value of each of the three parameters (α/ml^2), $\phi(0)$, and $\phi'(0)$. But a reasonably good understanding of this system can be gained only after studying the motions resulting from many combinations of these parameters. (It is to help with this problem that the change of variable $t = \sqrt{(l/g)}\ \tau$ was introduced, reducing the number of parameters in the system by one.) By contrast, one can examine, with much less effort, the analytically derived expressions on p. 67 and gain a fairly complete understanding of the roles of the several parameters in the spring-mass system. Thus it would be foolish to study that system by means of numerical integration.

Problems

11-46 Minimize the number of parameters in the differential equation resulting from Problem 11-24 by rewriting it in terms of dimensionless variables. Write and run a computer program to numerically integrate the equation.

11-47 Verify that the spring-mass-dashpot system, as idealized in the sketch, is governed by the differential equation

$$m\ddot{x} + c\dot{x} + kx = F_0 \cos \Omega t$$

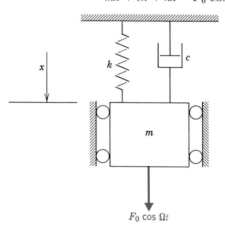

$$F_0 \cos \Omega t$$

Although this does not fall into one of the categories discussed in Section 11-3, it can be solved analytically by means of methods discussed in Chapter 18. The solution is

$$x = C e^{-\zeta \omega_0 t} \cos (\omega t + \theta) + X \cos (\Omega t - \phi)$$

where C and θ are arbitrary constants, and

$$\zeta = \frac{c}{2\sqrt{mk}} \qquad \omega_0 = \sqrt{\frac{k}{m}} \qquad \omega = \sqrt{1 - \zeta^2}\, \omega_0$$

$$X = \frac{F_0/k}{\sqrt{\left(1 - \frac{\Omega^2}{\omega_0^2}\right)^2 + \left(2\zeta \frac{\Omega}{\omega_0}\right)^2}}$$

$$\phi = \tan^{-1} \frac{2\zeta \frac{\Omega}{\omega_0}}{1 - \frac{\Omega^2}{\omega_0^2}}$$

(a) Rewrite the differential equation in terms of dimensionless variables so as to reduce the number of parameters in the system to as few as possible.

(b) Write and run a computer program that will numerically integrate the equation and compare values of x with the analytical solution.

11-48 Write and run a computer program to numerically integrate the differential equation of the Van der Pol oscillator,

$$\ddot{x} - \alpha(1 - x^2)\dot{x} + x = 0$$

For several different sets of initial conditions, carry out the integration until the motion is essentially periodic. Do this for $\alpha = 0.1$ and $\alpha = 1.0$.

11-49 On p. 110 the following differential equation is derived.

$$\ddot{\phi} + \left(\frac{g}{l} - \frac{a_0\omega^2}{l}\cos\omega t\right)\sin\phi = 0$$

It governs the parametrically excited pendulum depicted in Figure 12-5. Minimize the number of parameters by nondimensionalizing and write a computer program to numerically integrate the equation. Among others, try some small initial values for parameters in the regions

$$\left(\frac{a_0}{l}, \omega \sqrt{\frac{l}{g}}\right) = (0.05, 2) \text{ and } (0.5, 1)$$

11-5

INTEGRATED FORMS OF NEWTON'S SECOND LAW. Two of the integration procedures outlined in Section 11-3 can be extended beyond one-dimensional problems. The resulting integrals are important enough to warrant special terminology. For problems in which they can be readily evaluated, these integrated forms of Newton's second law are a step nearer the solution than the second law itself.

The Impulse-Momentum Relationship. If each member of the equation $\mathbf{f} = m\dot{\mathbf{v}}$ is integrated with respect to time

$$\int_{t_1}^{t_2} \mathbf{f}(t)\, dt = m\mathbf{v}_2 - m\mathbf{v}_1 \tag{11-7}$$

results.

The vector quantity on the left is called the *impulse* of the force $\mathbf{f}$ and is denoted as

$$\mathbf{f}^*_{1-2} = \int_{t_1}^{t_2} \mathbf{f}(t)\, dt \tag{11-8}$$

The vector quantity

$$\mathbf{p} = m\mathbf{v} \qquad (11\text{-}9)$$

is called the *momentum* of the particle. In words, then, Equation 11-7 states that the impulse of the force is equal to the change in the momentum of the particle. Whenever the force is given as an explicit function of time, the impulse may be readily calculated, and Equation 11-7 maybe used to relate it to the velocity change.

Example

A 35 000-lb aircraft is flying along a straight path. When traveling at a constant 800 mph, the pilot ignites the afterburner, inducing an added 12 000-lb thrust. Assuming the aerodynamic forces remain constant, how long must the afterburner fire in order that the aircraft achieve a speed of 1200 mph?

The magnitude and direction of thrust being constant, the impulse of the resultant force on the aircraft is in the direction of motion and has the magnitude

$$f^* = (12\ 000\ \text{lb})t$$

The change in momentum has the magnitude

$$|m\mathbf{v}_2 - m\mathbf{v}_1| = \left(\frac{35\ 000\ \text{lb}}{32.2\ \text{ft/s}_2}\right)(400\ \text{mi/hr})\left(\frac{5280\ \text{ft/mi}}{3600\ \text{s/hr}}\right)$$

$$= 638\ 000\ \text{lb·s}$$

The impulse-momentum relationship then gives us

$$t = \frac{638\ 000\ \text{lb·s}}{12\ 000\ \text{lb}} = 53\ \text{s}$$

In general, the impulse-momentum relationship is useful where something is known of the force as a function of time, and either the given information or the answer sought involves the time interval.

The Work-Kinetic Energy Relationship.   If each member of Equation 1-1 expressing Newton's second law is dot multiplied by the velocity of the particle, there results

$$\mathbf{f} \cdot \mathbf{v} = m\dot{\mathbf{v}} \cdot \mathbf{v} = \frac{d}{dt}\left(\frac{1}{2}\,m\mathbf{v} \cdot \mathbf{v}\right)$$

Integration with respect to time yields

$$\int_1^2 \mathbf{f} \cdot \frac{d\mathbf{r}}{dt}\,dt = \frac{mv^2}{2}\,\bigg|_1^2$$

Or,

$$\int_1^2 \mathbf{f} \cdot d\mathbf{r} = \frac{mv_2{}^2}{2} - \frac{mv_1{}^2}{2}$$ (11-10)

The scalar quantity on the left is called the *work* done by the force $\mathbf{f}$ as the particle moves from position $\mathbf{r}_1$ to $\mathbf{r}_2$, and it is denoted as

$$W_{1-2} = \int_1^2 \mathbf{f} \cdot d\mathbf{r}$$ (11-11)

The scalar quantity

$$T = \tfrac{1}{2} mv^2$$ (11-12)

is called the *kinetic energy* of the particle. In words, then, Equation 11-10 states that the work done on the particle as it moves from position $\mathbf{r}_1$ to position $\mathbf{r}_2$ is equal to its change in kinetic energy. Note that time has been eliminated from the force-acceleration relationship. For this reason the work-kinetic energy relationship is particularly useful in dynamics problems where neither the given information nor the answer sought involves the time interval.

As an example, let us determine the distance that the aircraft of the previous example travels during the acceleration stage. The work-kinetic energy relationship becomes, for this case,

$$12\ 000\ \text{lb}\ |\mathbf{r}_2 - \mathbf{r}_1| = \frac{35\ 000\ \text{lb}}{2(32.2\ \text{ft/s}^2)}\ [(1760\ \text{ft/s})^2 - (1174\ \text{ft/s})^2]$$
$$= 9.35 \times 10^8\ \text{ft-lb}$$

The distance traveled is thus

$$\frac{9.35 \times 10^8\ \text{ft-lb}}{12\ 000\ \text{lb}} = 78\ 000\ \text{ft} = 14.8\ \text{mi}$$

Observe that this computation in no way involved the time interval. As a check, and to illustrate the advantage of the work-kinetic energy relationship, you may wish to write the distance as a function of time and substitute the 53 s resulting from the first calculation.

As a final observation, note that both the impulse-momentum relationship and the work-kinetic energy relationship were developed from the form of Newton's second law, incorporating the acceleration observed from an inertial reference frame. Momentum, work, and kinetic energy must be used with care in any analysis involving a moving frame of reference, since these all include kinematic quantities that depend on the frame of reference.

Potential Energy. Whenever there is a force defined for each point throughout some region of space (such as the force that acts on a particle anywhere within the influence of the earth's attraction), we refer to the collection of values as a *force field*. For many force fields, the work done between two points in space,

$$W_{1-2} = \int_{\mathbf{r}_1}^{\mathbf{r}_2} \mathbf{f} \cdot d\mathbf{r}$$

is independent of the path along which the displacement increments are taken. When this is the case, the force field is said to be *conservative*, and we find it useful for analysis to define a scalar *potential energy function V*, as

$$V(\mathbf{r}) = -\int_{\mathbf{r}_0}^{\mathbf{r}} \mathbf{f}(\boldsymbol{\rho}) \cdot d\boldsymbol{\rho} \tag{11-13}$$

This is the negative of the work done by the force field on a particle as the

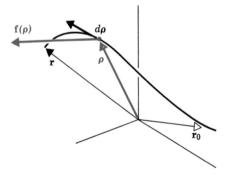

particle moves from position $\mathbf{r}_0$ to position $\mathbf{r}$. Or, equivalently, it is the work done *against* the force field by a force that would be required to hold the particle in static equilibrium as the particle is moved slowly from position $\mathbf{r}_0$ to position $\mathbf{r}$.

As an example, consider the gravitational force that the earth exerts on a particle. The work done by this force field on the particle as it moves from the surface of the earth to a height r (measured from the center of the earth), is

$$W = \int_{\mathbf{r}_0}^{\mathbf{r}} -\frac{mga^2}{\rho^2}\, \mathbf{u}_\rho \cdot d\boldsymbol{\rho} = -\int_a^r \frac{mga^2}{\rho^2}\, d\rho = mga^2 \left(\frac{1}{r} - \frac{1}{a}\right)$$

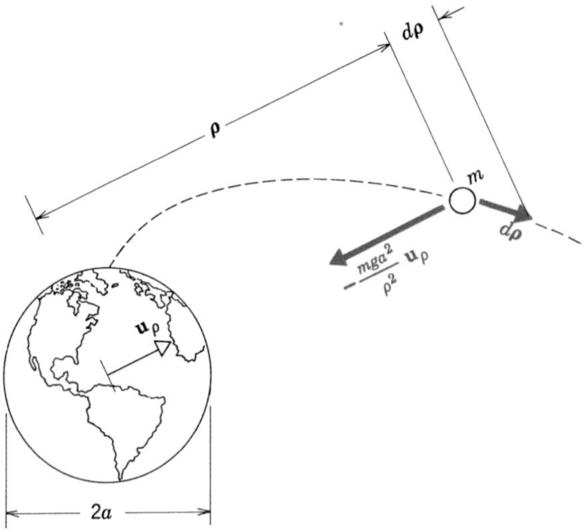

Since this work depends on the final position but not in any way on the path followed by the particle, this force field is conservative, and the potential is given as the negative of the work just computed:

$$V(r) = mga^2 \left(\frac{1}{a} - \frac{1}{r}\right) \tag{11-14a}$$

Observe that in terms of the height h above the surface of the earth, this becomes

$$V = mga^2 \left(\frac{1}{a} - \frac{1}{a+h}\right)$$

$$= mgh\, \frac{1}{1+\dfrac{h}{a}}$$

and when $h \ll a$,

$$V \approx mgh \tag{11-14b}$$

This last approximation is consistent with the assumption that the magnitude of the force is constant, equal to mg.

The choice of the lower limit $\mathbf{r}_0$ in the definition of potential (11-13) affects V only by an additive constant. The choice of this constant has no

influence on any of the relationships in which we use the potential function. To illustrate, let us compute the potential energy in a stretched
spring of stiffness k, using two different positions for the lower limit of integration.

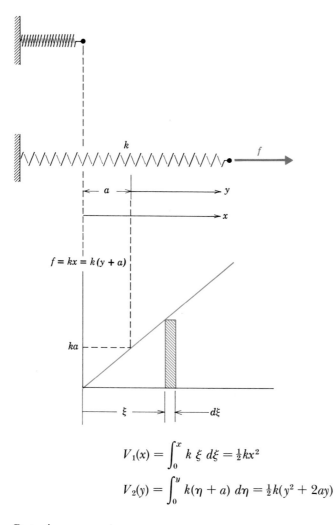

$$V_1(x) = \int_0^x k\,\xi\,d\xi = \tfrac{1}{2}kx^2$$

$$V_2(y) = \int_0^y k(\eta + a)\,d\eta = \tfrac{1}{2}k(y^2 + 2ay)$$

But, since $x = y + a$,

$$V_1 = \tfrac{1}{2}k(y + a)^2 = \tfrac{1}{2}k(y^2 + 2ay + a^2)$$
$$= V_2 + \tfrac{1}{2}ka^2$$

The two values of potential differ only by the additive constant $\tfrac{1}{2}ka^2$, the difference in potential between the two points chosen as the two lower limits.

It should be understood that not all position-dependent forces are conservative. Suppose, for example, a rocket is acted on by a thrust **f** that is controlled to depend on the position of the rocket as follows.

$$\mathbf{f} = \frac{F_0}{r_0} \left[(x + y)\, \mathbf{u}_x + (y - x)\, \mathbf{u}_y \right]$$

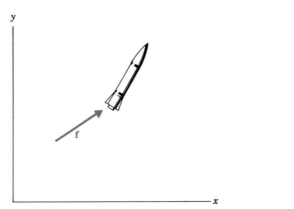

Computation of the work done by this force, along two different paths connecting the origin and the point (x,y), verifies that it is not conservative.

$$W_1 = \frac{F_0}{r_0} \int_0^x \xi\, d\xi + \frac{F_0}{r_0} \int_0^y (\eta - x)\, d\eta$$

$$= \frac{F_0}{2r_0} (x - y)^2$$

$$W_2 = \frac{F_0}{r_0} \int_0^y \eta\, d\eta + \frac{F_0}{r_0} \int_0^x (\xi + y)\, d\xi$$

$$= \frac{F_0}{2r_0} (x + y)^2$$

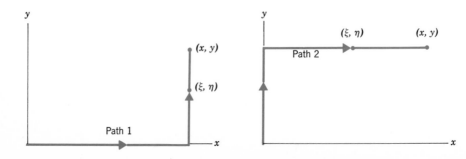

The reader who is familiar with the rudiments of the analysis of vector fields will recognize that the necessary and sufficient condition that a force field $\mathbf{f(r)}$ be conservative is that curl $(\mathbf{f})$ vanish throughout the region in question. This condition is expressed in rectangular Cartesian coordinates as

$$\frac{\partial f_x}{\partial y} - \frac{\partial f_y}{\partial x} = \frac{\partial f_y}{\partial z} - \frac{\partial f_z}{\partial y} = \frac{\partial f_z}{\partial x} - \frac{\partial f_x}{\partial z} = 0 \qquad (11\text{-}15)$$

Total Mechanical Energy. If the motion of a particle is such that the only forces that do work are conservative, the work can be expressed in terms of the potential function:

$$W_{1-2} = \int_{1}^{2} \mathbf{f} \cdot d\mathbf{r} = -(V_2 - V_1)$$

and the work-kinetic energy relationship (11-10) may be written as

$$-(V_2 - V_1) = T_2 - T_1$$

Or,

$$\boxed{T_2 + V_2 = T_1 + V_1} \qquad (11\text{-}16)$$

That is, the *total mechanical energy,*

$$E = T + V$$

remains constant.

You will find it very instructive to review the integration procedures developed on pp. 66–67 in light of these developments of kinetic and potential energies.

Example

The soapbox racer shown in Figure 11-4 is released from rest at the top of the hill. How fast will it be traveling when it reaches the finish line?

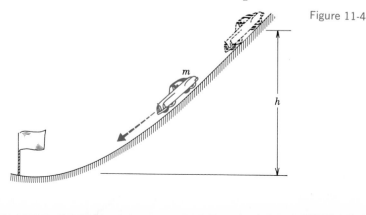

Figure 11-4

Assuming that friction is negligible, the reaction from the track is perpendicular to the local displacement increment, and hence does no work on the racer. With aerodynamic forces also neglected, the remaining force, that of gravity, is conservative. With reference to (11-14) and (11-16), the energy relationship may be written as

$$\tfrac{1}{2}mv^2 + 0 = \tfrac{1}{2}m(0)^2 + mgh$$

from which

$$v = \sqrt{2gh}$$

Observe that the speed achieved does not depend on the contour of the hill, even though the time of travel would be affected by this contour!

Problems

11-50   Using the work-kinetic energy relationship, we found that the aircraft in the example on p. 80 traveled 14.8 mi while accelerating from 800 to 1200 mph. Verify, by integrating the velocity with respect to time, that the 53-s interval, computed from the impulse-momentum relationship, yields the same distance.

11-51   An airplane is banking at constant altitude, at uniform speed and in a path of fixed curvature $1/a$.

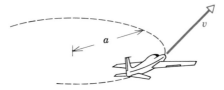

Using its *definition,* calculate the impulse of all external forces acting on the aircraft during a 90° turn. Check the answer by means of Equation 11-8.

11-52   Compute the work done in raising a body slowly from the surface of the earth to an altitude of 100 km, by:
(a) Assuming that the force has the constant value $f_g = mg$.
(b) By assuming the force varies according to Newton's law of gravitation, $f_g = mg\ a^2/r^2$.
What is the fraction of the error in (a) as compared with (b)?

11-53   Evaluate the work done in stretching the spring assembly shown a distance of 1 ft. Initial tension is 100 lb.

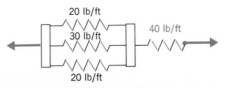

20 lb/ft

30 lb/ft

40 lb/ft

20 lb/ft

11-54 Determine the potential energy of the system as a function of displacement x. Assume the gas in the cylinder obeys the ideal gas law

$$p \, v^\gamma = \text{constant}$$

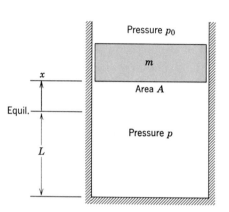

11-55 An object is dropped from height h onto the cushioning spring. What will be the deflection and maximum force in the spring?

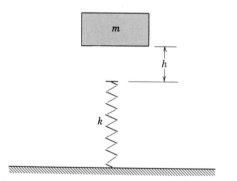

11-56 The pin of mass m is released from rest in the position shown where the spring of constant k is compressed to one half its free length l. Estimate the maximum velocity v reached by the pin, assuming friction effects to be negligible.

11-57 A 4.5-Mg aircraft lands on an aircraft carrier at a relative velocity of 27 m/s parallel to the deck. It is immediately caught by an arrester gear that exerts a decelerating force directly proportional to the travel of the aircraft, the constant of proportionality being 1460 N/m. When its speed has been reduced to 6.7 m/s, the arrester releases the aircraft and the pilot brings it to a stop with a constant 4.45 kN braking force. What is the total landing distance measured from the point of contact with the arrester gear?

11-58 A 12-ton aircraft is to be launched at 95 mph from a catapult that has a carriage run of 55 ft. The launching mechanism is such that the force varies linearly with displacement of the carriage from an initial value f to a final value $f/2$. Evaluate f.

11-59 An airplane of mass m is descending with a vertical velocity v_A; it meets the deck of a carrier that has, at that time, a downward velocity v_C. The airplane's shock absorber is a spring with a restoring force $f_s = kx$. To what maximum deflection x_{max} will the spring deflect, and what will be the maximum spring force if the aerodynamic lift remains equal to the gravitational force during this short time?

11-60 A sheriff fires a pistol at the rear tire of a getaway car, both cars traveling at 141 ft/s. From a stationary position the pistol produces a muzzle velocity of 729 ft/s in the 200-grain bullet, so that its kinetic energy is 236 ft·lbf. As a moving observer, the sheriff claims that the bullet will have a speed relative to the ground of $v = 141$ ft/s $+ 729$ ft/s $= 870$ ft/s. An engineer standing on the sidewalk disagrees, saying that since there is a fixed amount of propellant in the cartidge, the change in kinetic energy of the bullet will be 236 ft·lbf so that

$$\tfrac{1}{2}mv^2 - \tfrac{1}{2}m \ (141 \ \text{ft/s})^2 = 236 \ \text{ft·lb}$$

This yields $v = 743$ ft/s. Which (if either) is correct, and why is the other (or both) incorrect?

11-61 A tennis ball, traveling at speed v, is met by a tennis racquet traveling at speed V. Assuming the racquet maintains its speed during contact with the ball,
(a) What is the maximum contact force?
(b) If there is no energy dissipation in the deformation-rebound mechanism, what is the impulse applied to the ball?
(c) How much work is done on the ball?

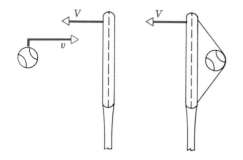

11-62   The 5-kg box, with an initial downward speed of 3 m/s, slides down the ramp, under the influence of gravity and the Coulomb friction coefficient $\mu = 0.27$.

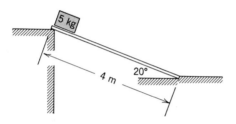

 (a) Using the work-energy idea, determine the speed of the box when it reaches the bottom.

 (b) Considering the force on the *ramp*, note that the friction force appears to do work on it, yet the ramp experiences no change in kinetic energy. Explain carefully. (*Suggestion.* Look closely at the definition of work.)

11-63   The box shown slides down the inclined plane, contacts the spring, compresses it, and then rebounds. The coefficient of friction between the box and the plane is μ, the angle between the plane and the horizontal is θ, the spring constant is k, and the mass of the box is m. Find the height to which it rebounds.

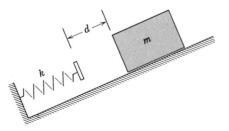

11-64   The unstretched length of the spring is $\frac{1}{2}$ ft. Estimate the velocity

v which the block A will strike the support at B assuming friction to be negligible.

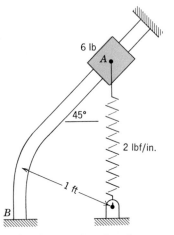

6 lb

A

45°

2 lbf/in.

1 ft

B

11-65 Referring to the sketch on p. 82, is $d\rho$ equal to $d|\rho|$, $|d\rho|$, both of these, or neither?

11-66 Estimate the escape velocity for a vehicle launched from the surface of the moon. The radius of the moon is 1080 mi and the acceleration due to gravity on the surface of the moon is 5.31 ft/s². If atmospheric resistance is neglected, what would be the escape velocity of a vehicle launched from the surface of the earth? (*Suggestion.* If the initial velocity is the escape velocity, then $v \to 0$ as $r \to \infty$.)

11-67 The block of mass m is released from rest at the position defined by X_1. The coefficient of friction between the block and the floor is μ.
(a) What is the potential energy in the spring before the block is released?
(b) What is the distance X_2 for which the block first comes to rest?
(c) Under what condition will the block *remain* in this position?

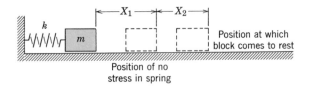

X_1 | X_2

k

m

Position at which block comes to rest

Position of no stress in spring

11-68 The force exerted by a particle carrying an electric charge e_1 on another particle carrying an electric charge e_2 is found to be equal to $(e_1 e_2/r^2)$, where r is the distance between the particles. Derive a potential function for the two-particle system. Evaluate the work required to move the particle of charge e_2 from r_A to r_B.

11-69   The forces acting on the particle shown are functions of position

$$f_x = Fy$$
$$f_y = Fx + Gy^2$$

(a) Does a potential function exist such that f_x and f_y are derivable from it? If so, what is it?
(b) Evaluate the work done as the particle moves from (1, 1) to (3, 6) along the straight line segments (1, 1) to (3, 1) to (3, 6).
(c) Evaluate the work done as the particle moves from (1, 1) to (3, 6) along the straight line segments (1, 1) to (1, 6) to (3, 6).

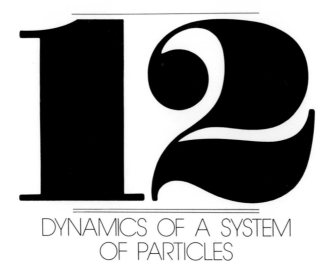

DYNAMICS OF A SYSTEM
OF PARTICLES

A *system of particles* is a collection of material of fixed identity.* While each particle within a system may be analyzed by laws that we studied in the previous chapter, such detailed analysis can easily become so complex as to be impossible to complete. In many cases such detail is, fortunately, unnecessary. Knowledge of certain overall aspects of the motion of the system may fulfill our needs and be practicably attained. A study of laws that govern these overall aspects will be the subject of this chapter.

12–1

MOTION OF THE CENTER OF MASS. Let us resolve the resultant force $\mathbf{f}_i$ acting on the ith particle in the system into the resultant force $\mathbf{f}_{ie}$ that results from interaction with bodies external to the system and the forces that result from interaction with the other particles within the system. That is, denoting by $\mathbf{f}_{i/j}$ the force acting on the ith particle due to

---

* A *system* may also be defined as the collection of material within a prescribed boundary enclosure, with the possibility that material can enter and leave the system during a time interval. This possibility is excluded in the meaning of the term *system of particles*.

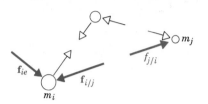

the jth particle, we have

$$\mathbf{f}_i = \mathbf{f}_{ie} + \sum_j \mathbf{f}_{i/j}$$

Then, with the mass of the ith particle denoted by m_i and its inertial-observed acceleration by $\mathbf{a}_i$, Newton's second law for this particle may be written as

$$\mathbf{f}_{ie} + \sum_j \mathbf{f}_{i/j} = m_i \, \mathbf{a}_i \qquad (i = 1,2,3, \ . \ . \ . \ ,n) \tag{12-1}$$

The note $(i = 1,2, \ . \ . \ . \ ,n)$ indicates that the law applies to *each* of the n particles within the system. If the n equations are added, there results

$$\sum_i \mathbf{f}_{ie} + \sum_i \sum_j \mathbf{f}_{i/j} = \sum_i m_i \, \mathbf{a}_i$$

Now, Newton's third law states that $\mathbf{f}_{j/i} = -\mathbf{f}_{i/j}$, so that the sum

$$\sum_i \sum_j \mathbf{f}_{i/j} = \mathbf{0}.$$

That is, in adding all the forces, the internal forces, which occur in equal magnitude, oppositely directed pairs, are self-canceling. Denoting the resultant of the external forces by $\mathbf{f}$, we then have

$$\mathbf{f} = \sum_i m_i \, \mathbf{a}_i \tag{12-2a}$$

The position vector $\mathbf{c}$, defined by the relationship

$$\sum_i m_i \mathbf{r}_i = \left(\sum_i m_i \right) \mathbf{c} \tag{12-3a}$$

locates a point C, called the *center of mass* of the system. It is instructive to verify that the point C has no dependence on the choice of the reference point O from which the position vectors $\mathbf{r}_i$ emanate, and that it is the one point that has the property

$$\sum_i m_i \boldsymbol{\rho}_i = \mathbf{0}$$

where $\boldsymbol{\rho}_i$ is the position vector from C to the ith particle. With point O

fixed in an inertial reference frame, differentiation of (12-3a) yields

$$\sum_i m_i \, \mathbf{v}_i = m \, \mathbf{v}_C \qquad (12\text{-}3b)$$

and a successive differentiation yields

$$\sum_i m_i \, \mathbf{a}_i = m \, \mathbf{a}_C \qquad (12\text{-}3c)$$

where $m = \Sigma m_i$ is the total mass in the system. With the help of (12-3b) and (12-3c), Equation 12-2a can also be written as

$$\mathbf{f} = \frac{d}{dt}\,(m\,\mathbf{v}_C) = m\,\mathbf{a}_C \qquad (12\text{-}2b)$$

Thus the *resultant external force influences the motion of the mass center in exactly the same way a force acting on a single particle influences the motion of the particle.* This is the very useful "principle of motion of the center of mass."

Example

The rigid, 620-lb rotor shown in Figure 12-1a is spinning in solidly mounted bearings at a constant rate of 1720 rpm. Measured lateral reac-

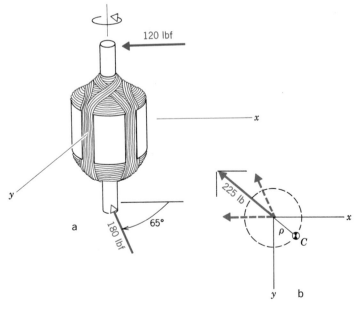

120 lbf

225 lb

180 lbf

65°

a

b

Figure 12-1

tions of 120 and 180 lbf from the bearing supports are as shown. The force vectors rotate with the rotor. Determine the distance from the spin axis to the center of mass.

The resultant lateral force is as shown in Figure 12-1b. The mass center is moving in a circular path around the spin axis at 1720 rpm; therefore its acceleration, directed toward the spin axis, has the magnitude

$$a_C = \rho \left[\frac{(1720 \text{ rpm})(2\pi \text{ rad/rev})}{60 \text{ s/min}} \right]^2$$
$$= (32\ 440 \text{ rad}^2/\text{s}^2)\rho$$

in which ρ is the desired distance. According to Equation 12-3c,

$$255 \text{ lbf} = \frac{620 \text{ lbf}}{386 \text{ in/s}^2} (32\ 440 \text{ s}^{-2}) \rho$$

from which

$$\rho = 0.0049 \text{ in}$$
$$= 0.124 \text{ mm}$$

The *impulse-momentum relationship for a system of particles* is obtained in the same manner as the analogous relationship for a single particle. We define the total linear momentum of the system as

$$\mathbf{p} = \sum_i (m_i \mathbf{v}_i) \qquad (12\text{-}4)$$

This and Equation 12-3b provide us with still another way to express the force-motion relationship.

$$\boxed{\mathbf{f} = \dot{\mathbf{p}}} \qquad (12\text{-}2c)$$

Integration with respect to time yields

$$\boxed{\int_{t_1}^{t_2} \mathbf{f} \, dt = \mathbf{p}_2 - \mathbf{p}_1} \qquad (12\text{-}5)$$

That is, the impulse of the resultant external force is equal to the change in the total momentum of the system.

Example

A moving railroad car strikes another car that is stationary, the cars becoming coupled on contact. What will be their speed v_1 after impact?

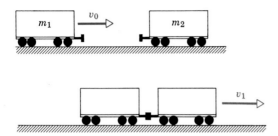

The resultant external force acting on the system consisting of the two cars is zero. Equation 12-5 then reduces to

$$m_1 v_0 + m_2(0) = (m_1 + m_2)\, v_1$$

or

$$v_1 = \frac{m_1}{m_1 + m_2}\, v_0$$

As another example, consider the action of a rocket that is propelled by the expulsion of gas. The exit velocity of the gas relative to the rocket, $\mathbf{v}_E$, is normally known. At time t, the mass of the rocket and fuel is $m(t)$,

$\mathbf{v}_E$ Relative to rocket

and its momentum is $m(t)\mathbf{v}(t)$. During a short time increment Δt, the mass $m(t) - m(t + \Delta t)$ is expelled from the rocket, changing its velocity to $\mathbf{v}(t + \Delta t) + \mathbf{v}_E$. The impulse-momentum relationship, applied to the system consisting of mass $m(t)$, becomes

$$\mathbf{f}\, \Delta t = m(t + \Delta t)\mathbf{v}(t + \Delta t) + [m(t) - m(t + \Delta t)][\mathbf{v}(t + \Delta t) + \mathbf{v}_E]$$
$$- m(t)\mathbf{v}(t)$$

Division by Δt and simplification result in

$$\mathbf{f} = m(t) \frac{\mathbf{v}(t + \Delta t) - \mathbf{v}(t)}{\Delta t} - \frac{m(t + \Delta t) - m(t)}{\Delta t} \mathbf{v}_E$$

Or, as $\Delta t \to 0$,

$$\mathbf{f} = m \frac{d\mathbf{v}}{dt} - \frac{dm}{dt} \mathbf{v}_E \tag{12-6}$$

Observe that $-(dm/dt)$ is the mass flow rate from the rocket.

Suppose the rocket has an initial mass m_0, expels gas at a constant rate μ and at constant relative exit speed v_E. If, further, it is fired vertically with gravity the only external force, the vectors in Equation 12-6 are all directed vertically, and the component equivalent of this equation is

$$-mg = m \frac{dv}{dt} + \frac{dm}{dt} v_E$$

The rearrangement

$$dv = -v_E \frac{dm}{m} - g \, dt$$

permits the first integration.

$$v = -v_E \log m - gt + C_1$$

When $t = 0$, $v = 0$ and $m = m_0$, so that $C_1 = v_E \log m_0$. Insertion of this value results in

$$v = v_E \log \frac{m_0}{m} - gt$$

With $m(t) = m_0 - \mu t$, a further integration gives the height as

$$z = \frac{v_E m_0}{\mu} \left[\frac{\mu t}{m_0} - \left(1 - \frac{\mu t}{m_0}\right) \log \frac{1}{1 - \mu t/m_0} \right] - \frac{1}{2} g t^2$$

Observe that in the law (12-2b) and the integrated form (12-5), which governs the motion of the center of mass, we have no information con-

Figure 12-2

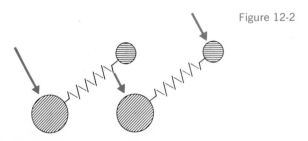

cerning any motion of various parts of the system relative to the mass center. For example, Equation 12-2b applied to the system of Figure 12-2 under the two different sets of forces will give the same result for each. While the detailed motion of the system will differ, the effect of the forces on the motion *of the mass center* will be identical in each case. Consideration of "overall rotational motion," in the next section, will give enough further detail to distinguish between the two cases of Figure 12-2; however, much detail of the general motion of an *n*-particle system will remain hidden from these laws, too.

Problems

12-1   Add the equations resulting from Problem 8-3b and 8-3c. What basic law, derived in this chapter, is expressed by the result? What system could you isolate for this problem to give the tension between the engine and the first car as the only unknown in the corresponding equation?

12-2   A 180-lb man walks from one end of a 12-ft, 55-lb canoe to the other. Initially, the canoe is stationary, on still water. How far does the canoe move?

12-3   A flat car can roll without resistance along a horizontal track as shown. Initially, the car together with a man of mass m is moving to the right with speed v. What increment of velocity Δv will the car obtain if the man runs with speed u relative to the floor of the car and jumps off at the left?

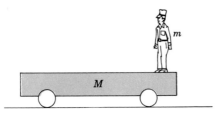

12-4   An aircraft of mass m is catapulted with a velocity v from an aircraft carrier of mass M. Just before the catapult is fired, the ship is moving with a velocity V. Assuming that the acceleration of the plane is due entirely to the action of the catapult, estimate the effect of the catapulting on the ship's velocity. How much of the plane's kinetic energy change during catapulting is supplied by the movement of the catapult relative to the ship?

12-5   A 1-Mg car is winched forward on the 8-Mg truck. The brakes of the truck have not been set prior to the winching process. Estimate the distance the truck will move during the process.

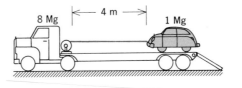

12-6 A 160-grain bullet leaves the muzzle of a rifle at 3150 ft/s. The barrel is 22 in. long and has an inside diameter of 0.283 in.

(a) What is the average (with respect to the bullet progression down the barrel) gas pressure behind the bullet?

(b) The rifle weighs 8.5 lb. Estimate its velocity toward the firer's shoulder as the bullet leaves the barrel.

(c) What is the average force exerted on the firer's shoulder, assuming it is pushed back 5 in.?

12-7 The block of mass m_1 is dropped from rest from the position indicated and slides freely down the rod until it strikes and becomes coupled with the lower block. Neglecting the mass of the spring and rod, determine the resulting extension in the spring.

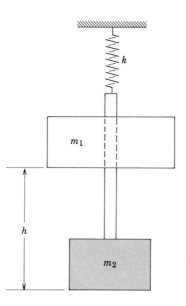

12-8 The bullet, traveling horizontally at speed v, strikes the ballistic pendulum and becomes imbedded. The mass M is much larger than the mass of the bullet. Write the relationship between the speed v and the angle θ, defining the maximum amplitude of swing of the pendulum.

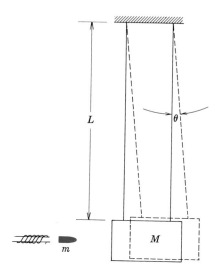

12-9 The ballistic pendulum of the previous problem is struck by a 130-grain bullet. The 25-lb block is suspended by 15-ft lines. A horizontal displacement of $18\frac{3}{4}$ inches is observed after impact. Estimate the velocity of the bullet.

12-10 A 135-kg cart is traveling at 0.5 m/s when a 45-kg sack drops vertically into the cart. The cart then strikes the bumper, which has a spring constant of 8.75 kN/m. What is the maximum force in the bumper?

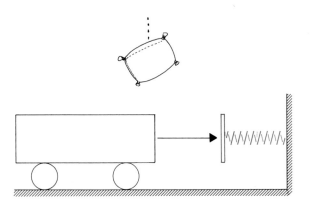

12-11 Gravel leaves the chute at velocity v_1 and at the rate of μ kg/s. The flat car has mass m_0 when empty, and rolls under the stream of gravel without friction and with no external forces applied. If time t is counted from the instant at which the front end of the car reaches the stream, what is the velocity $v(t)$ of the car during the time the car is under the stream? (*Suggestion.* Consider a system of particles consisting of the flatcar plus all the gravel that will be in the car up to time t.)

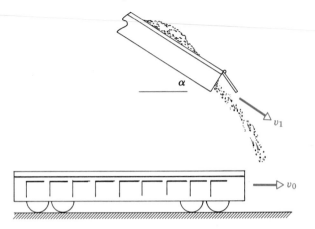

12-12 A passenger train consisting of 14 cars weighing 30 tons each and an engine weighing 70 tons are proceeding up a rise of 1 ft in 100 ft at 30 mph when the brakes on all but the last four cars lock, causing the wheels to slide. Assuming a coefficient of friction 0.30 between wheels and rails, estimate the time required for the train to be brought to a stop.

12-13 An electric motor is bolted to its foundation as shown. Due to inaccuracies in workmanship, the center of mass C of the rotor does not coincide with the axis of rotation O but describes a circle of radius e during motion. Determine the vertical component of the force exerted by the motor on the foundation.

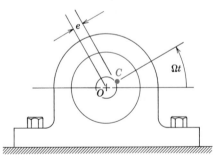

12-14 A rotating machine with an unbalance like that of the previous problem is mounted on a flexible support. The springs transmit a force proportional to displacement and the dashpot a force pro-

portional to velocity. The distance from the center of rotation to
the center of mass of the rotor is e, the angular velocity of the
rotor is Ω, the mass of the rotor is m, and the mass of the rotor
plus non-rotating parts is M. Derive the differential equation that
describes the vertical displacement $y(t)$.

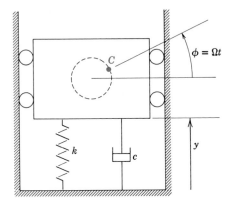

12-15 An aircraft of mass M hooks onto the end of an arresting chain of
mass m and length l that is strung out along the runway parallel to
the landing path of the aircraft. Neglecting friction between the
runway and moving objects, estimate the velocity and acceleration
of the aircraft as functions of distance traveled and as functions of
time. Sketch curves showing these relationships.

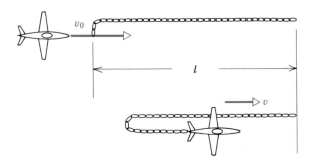

12-16 A rocket is fired vertically upward from an initial altitude of
15 km. Neglecting air resistance and assuming g is a constant
9.8 m/s², estimate the altitude the rocket will reach if its initial
mass is 3.6 Mg, its burning rate is 900 kg/s, its total burning time
is 3 s, and the exhaust velocity is 1800 m/s.

12-17 A rocket starts from rest with mass m_0 and falls under the influence of gravity. Gravitational effects are partly overcome by ex-

pelling mass at a rate μ and directly downward at speed v_E relative to the rocket. Determine the distance the rocket falls in time t.

12-18 Show that in the absence of any external forces, the total linear momentum of a group of particles remains unchanged, regardless of any internal interactions between particles.

12-19 Show that the mass center C, defined by Equation 12-3a, has no dependence on the choice of the reference point for the position vectors $\mathbf{r}_i$, and that $\Sigma\, m_i\, \boldsymbol{\rho}_i = \mathbf{0}$.

12-20 A 50-lb satellite is traveling at a speed of 25,000 ft/s. A small adjustment in its flight path is made by firing a 1-lb lead pellet from the satellite in a direction normal to its flight path. What ejection velocity is required to change the flight path direction by 0.01 rad? Assume the ejection time is instantaneous.

12-2

ROTATIONAL CONSIDERATIONS: MOMENTS OF FORCE AND MOMENTUM. Relative to the point O fixed in an inertial reference frame, the moment of external forces is defined as

$$\boxed{\mathbf{M}_O = \sum_i \mathbf{r}_i \times \mathbf{f}_{ie}} \qquad (12\text{-}7)$$

and the moment of momentum, or *angular momentum*, as

$$\boxed{\mathbf{H}_O = \sum_i \mathbf{r}_i \times \mathbf{p}_i} \qquad (12\text{-}8)$$

The relationship between these two quantities is developed much as was the force-linear momentum relationship (12-5). The force acting on the ith particle is resolved as before,

$$\mathbf{f}_i = \mathbf{f}_{ie} + \sum_j \mathbf{f}_{i/j}$$

and each member of Equation 12-1 cross-multiplied with the position vector $\mathbf{r}_i$:

$$\mathbf{r}_i \times \left(\mathbf{f}_{ie} + \sum_j \mathbf{f}_{i/j} \right) = \mathbf{r}_i \times m_i \dot{\mathbf{v}}_i = \frac{d}{dt}\,(\mathbf{r}_i \times m_i \mathbf{v}_i)$$

Addition of the individual equations governing each particle results in

$$\sum_i \mathbf{r}_i \times \mathbf{f}_{ie} + \sum_i \sum_j \mathbf{r}_i \times \mathbf{f}_{i/j} = \frac{d}{dt} \sum_i \mathbf{r}_i \times m_i \mathbf{v}_i$$

Now if, in addition to satisfying Newton's third law $\mathbf{f}_{i/j} = -\mathbf{f}_{j/i}$, each action-reaction pair of internal forces has a common line of action, then

$$\mathbf{r}_i \times \mathbf{f}_{i/j} = -\mathbf{r}_j \times \mathbf{f}_{j/i}$$

That is, the *moment* of $\mathbf{f}_{i/j}$ cancels the moment of $\mathbf{f}_{j/i}$. When this is the case,* we have

$$\sum_i \mathbf{r}_i \times \mathbf{f}_{ie} = \frac{d}{dt} \sum_i \mathbf{r}_i \times \mathbf{p}_i$$

Or,

$$\boxed{\mathbf{M}_O = \dot{\mathbf{H}}_O} \qquad (12\text{-}9)$$

Integration of Equation 12-9 with respect to time gives the *angular impulse-angular momentum* relationship

$$\int_{t_1}^{t_2} \mathbf{M}_O \, dt = \mathbf{H}_{O2} - \mathbf{H}_{O1} \qquad (12\text{-}10)$$

Example

A satellite orbiting the earth has a speed of 8700 m/s at the perigee of its orbit, where the altitude is 290 km. What is its speed at the apogee, where the altitude is 5087 km?

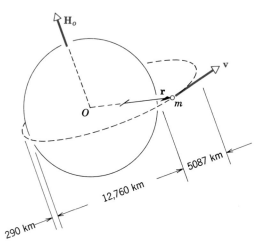

The moment about the center of the earth, of the force acting on the satellite, is zero. Then, according to Equation 12-10, the angular momentum about the same point is constant. The magnitude of $\mathbf{H}_O$ may be

---

* This is normally the case. A notable exception occurs in the electromagnetic interaction of charged particles.

computed in terms of the altitude and velocity at perigee, where the angle between $\mathbf{r}$ and $m\mathbf{v}$ is $\pi/2$; thus

$$H_O = [(6380 + 290) \text{ km}] \ m \ (8700 \text{ m/s})$$

where m is the mass of satellite. Since $\mathbf{H}_O$ is constant, its magnitude must be the same at apogee, where the angle between $\mathbf{r}$ and $m\mathbf{v}$ is again $\pi/2$.

$$H_O = [(6380 + 5087) \text{ km}] \ m \ v$$

Equating the two expressions for H_O leads to

$$v = \frac{6670}{11\ 467} \ (8700 \text{ m/s})$$
$$= 5060 \text{ m/s}$$

It is instructive to verify that the total mechanical energy is also conserved.

Example

The two particles shown in Figure 12-3a are held in a fixed position relative to the rotating shaft by a truss of negligible mass. The shaft is spinning at a constant rate ω in rigidly supported bearings. Evaluate the angular momentum about the mass center, $\mathbf{H}_C$, its rate of change $\dot{\mathbf{H}}_C$, and the lateral reactions at the bearings.

The velocities of the two particles are $2a\omega\mathbf{u}_y$ and $-a\omega\mathbf{u}_y$, respectively. The angular momentum about the mass center is then

$$\mathbf{H}_C = (2a\mathbf{u}_x + 2b\mathbf{u}_z) \times m(2a\omega\mathbf{u}_y)$$
$$+ (-a\mathbf{u}_x - b\mathbf{u}_z) \times 2m(-a\omega\mathbf{u}_y)$$
$$= 6ma\omega(a\mathbf{u}_z - b\mathbf{u}_x)$$

Since the magnitude and orientation of this vector remains fixed in the body, it swings around the z axis at the rate ω, as shown in Figure 12-3b. Observe that this is exactly the manner in which the vector $\mathbf{r}$ in Figure 10-10 varies in the fixed reference frame. The rate of change of this vector was shown to be $\dot{\mathbf{r}} = \boldsymbol{\omega} \times \mathbf{r}$ (Equation 10-10). Therefore we have, in this case,

$$\dot{\mathbf{H}}_C = \boldsymbol{\omega} \times \mathbf{H}_C$$
$$= \omega H_x \mathbf{u}_y$$
$$= -6mab\omega^2\mathbf{u}_y$$

Now, from Equation 12-9, we see that the reactions from the bearings must form a couple with moment equal to this value of $\dot{\mathbf{H}}_C$. Therefore

Figure 12-3

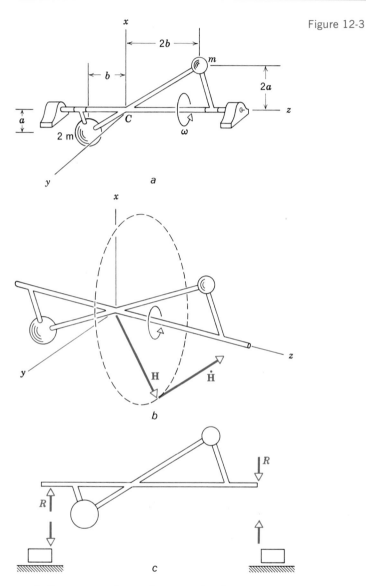

each force must have the magnitude

$$R = \frac{6mab\omega^2}{L}$$

where L is the distance between bearings, and act in the directions indicated in Figure 12-3c. These reactive forces rotate with the body; additional reactions induced by gravity have not been considered.

Example

The flyball governor shown in Figure 12-4 is spining around the vertical axis at the rate $\dot{\psi}$, while the arms swing at the rate $\dot{\chi}$. Neglecting mass

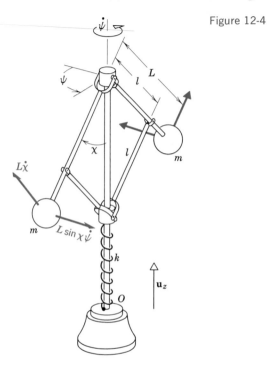

Figure 12-4

other than that of the balls, evaluate the angular momentum about the point O, in terms of m, L, χ, $\dot{\chi}$, and $\dot{\psi}$.

The contributions to angular momentum from the two velocity components $L\dot{\chi}$ (one for each ball) are self-canceling, because their directions oppose one another. However, each of the contributions from the two velocity components $L \sin \chi\dot{\psi}$ is directed upward. Therefore the angular momentum of the two-mass system, about O, is

$$\mathbf{H}_O = 2m(L \sin \chi)^2 \, \dot{\psi} \, \mathbf{u}_z$$

If the vertical shaft is free of torque, this quantity will remain constant; then the spin rate $\dot{\psi}$ will decrease and increase, respectively, as the angle χ increases and decreases.

Moments of Force and Momentum About a Moving Reference Point.   It sometimes happens that the point about which moments are most readily evaluated is moving. In view of this, let us generalize the preceding result

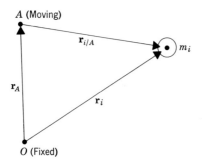

to the case where the reference point is moving in an inertial reference frame. For this case we write

$$\mathbf{r}_{i/A} \times \left(\mathbf{f}_{ie} + \sum_j \mathbf{f}_{ij} \right) = \mathbf{r}_{i/A} \times m_i \frac{d^2}{dt^2} (\mathbf{r}_{i/A} + \mathbf{r}_A)$$

and again add the equations, assuming that moments of the internal action-reaction pairs are self-canceling.

$$\sum_i \mathbf{r}_{i/A} \times \mathbf{f}_{ie} = \sum_i \mathbf{r}_{i/A} \times m_i \dot{\mathbf{v}}_{i/A} + \left(\sum_i m_i \mathbf{r}_{i/A} \right) \times \ddot{\mathbf{r}}_A$$

$$= \frac{d}{dt} \left(\sum_i \mathbf{r}_{i/A} \times m_i \mathbf{v}_{i/A} \right) + m\mathbf{r}_{C/A} \times \mathbf{a}_A$$

We define the angular momentum about the point A as*

$$\boxed{\mathbf{H}_A = \sum_i \mathbf{r}_{i/A} \times m_i \, \mathbf{v}_{i/A}} \tag{12-11}$$

This can be shown (see Problem 12-33) to be related to the angular momentum about the mass center C by

$$\mathbf{H}_A = \mathbf{H}_C + \mathbf{r}_{C/A} \times m\mathbf{v}_{C/A} \tag{12-12}$$

In terms of the definition (12-11), the above moment equation may be written as

$$\boxed{\mathbf{M}_A = \dot{\mathbf{H}}_A + \mathbf{r}_{C/A} \times m\mathbf{a}_A} \tag{12-13}$$

---

* Note that the velocity appearing in the definition of $\mathbf{H}_A$ is the velocity *difference* $\mathbf{v}_{i/A} = \mathbf{v}_i - \mathbf{v}_A$. A different quantity, $\mathscr{H} = \Sigma \, \mathbf{r}_{i/A} \times \mathbf{p}_i$, is sometimes taken as the angular momentum about the point A. The distinction might be described by using the terms "moment of relative momentum" and "moment of absolute momentum". The equation equivalent to (12-13) is

$$\mathbf{M}_A = \dot{\mathscr{H}}_A + \mathbf{v}_A \times \mathbf{p}$$

Example

The support point for the pendulum shown in Figure 12-5 is given a prescribed oscillatory displacement in the vertical direction. Derive the differential equation governing the plane motion $\phi(t)$.

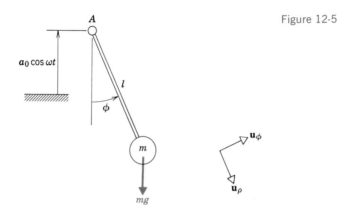

Figure 12-5

With the reference point A and the coordinate directions as shown, we have:

$$\mathbf{M}_A = -mgl \sin \phi \, \mathbf{u}_z$$

$$\mathbf{H}_A = (l)\,(m)\,(l\dot{\phi})\mathbf{u}_z$$

$$\mathbf{r}_{C/A} = l\,\mathbf{u}_\rho$$

$$\mathbf{a}_A = \omega^2 a_0 \cos \omega t(\mathbf{u}_\rho \cos \phi - \mathbf{u}_\phi \sin \phi)$$

Substitution of these values into Equation 12-13 results in

$$-mgl \sin \phi = ml^2\ddot{\phi} - ma_0l\omega^2 \cos \omega t \sin \phi$$

Or,

$$\ddot{\phi} + \left(\frac{g}{l} - \frac{a_0\omega^2}{l} \cos \omega t\right) \sin \phi = 0$$

With the simplification $\sin \phi = \phi$, valid for small ϕ, this differential equation is called Mathieu's equation. Its solutions and their properties have been studied extensively. It predicts some rather strange behavior; perhaps the most striking characteristic is that for certain combinations of amplitude and frequency of support motion, the upside-down equilibrium position, $\phi = \pi$, is *stable!*

Problems

12-21   Explain how the equations of moment equilibrium, used so extensively in statics, are a consequence of Newton's second and third laws.

12-22   Explain how a skater increases his rotation rate while performing a vertical spin in a stationary location on the ice.

12-23   Evaluate the angular momentum contribution from *one* of the flyballs in Figure 12-4.

12-24   A particle, guided to slide frictionlessly along a uniformly rotating tube, can be shown to have motion given by

$$s(t) = s_0 \cosh \Omega t$$

where s_0 is the position from which the particle is released from relative rest.

(a) Evaluate the reaction that the mass exerts on the tube, showing its magnitude and direction.

(b) How much work does the tube do on the particle during the first revolution (assuming the tube is long enough that the particle remains within it)?

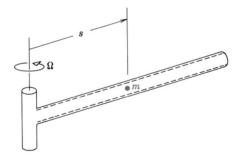

12-25   The particle and tube of the previous problem are now rotating about the vertical free of any external moment about the vertical spin axis. The mass of the tube is M, its length is L. Derive the differential equations that describe the angular velocity $\Omega(t)$ and the distance $s(t)$ locating the particle. [*Suggestion.* Equation 10-7 will be helpful for evaluating the velocity and acceleration of the particle. (12-9) may also be helpful.]

12-26   (a) Work Problem 12-27 for the special case in which $b = a$. Apply Equation 12-2 instead of Equation 12-9 to the separated system.

(b) Estimate the time required under no-load conditions to get the drive system up to its operating speed of 62.8 rad/s. The mass of the sprockets is negligible compared with the 6.5-kg mass of the chain. A constant torque of 80 N·m is applied to the drive sprocket, which has a radius of 200 mm.

12-27   The mass of the drive chain is m, and the angular velocity of the drive sprocket is $\omega(t)$. A couple of moment M is applied to the drive sprocket.

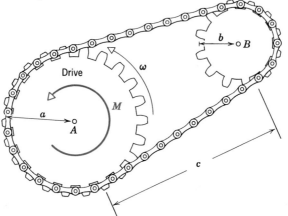

(a) Evaluate the angular momentum of the chain about A, the center of the drive sprocket. What is the angular momentum of the chain about B?

(b) Neglecting inertia of the sprockets, apply Equations 12-2 and 12-9 to the system consisting of the chain and sprockets. Apply Equation 12-9 to the system consisting of the driven sprocket and a portion of the chain isolated by separations through the straight portions of the chain.

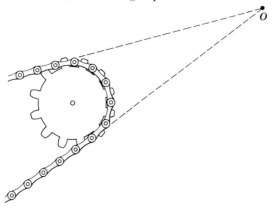

(c) What is the relationship between M and the angular acceleration of the drive sprocket?

12-28　A speed-dependent torque, obeying

$$T = 160 \text{ N·m} \left(1 - \frac{\omega}{62.8 \text{ rad/s}}\right)$$

is applied to the drive sprocket of Problem 12-26. Estimate the time required to reach 98% of operating speed in this case.

12-29　A 35-lb turkey walks on a 20-lb conveyor belt, starting from point A. The inertia and friction of the rollers is negligible. The system is initially at rest. Determine how far the turkey must walk to reach the other roller, which is 8 ft away from A.

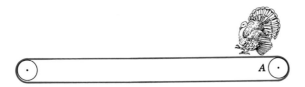

12-30　A 1.5-kg ball on a 1-m long cord is spinning about a vertical axis at 40 rpm when struck by a 10 g bullet traveling as shown at 800 m/s in the plane of rotation of the ball-cord system. Estimate the angular velocity of the ball-cord system after the bullet is embedded in the ball.

12-31　The monkey drops from his perch to swing to the tree on the other side of the clearing. When the vine becomes taut, the monkey cushions the blow by flexing his arms. The mass of the vine is negligible.

(a) Evaluate the velocity of the monkey immediately before and immediately after the impact of the vine straightening.

(b) How much energy has the monkey absorbed in his arms?

(c) Will he escape the lion?

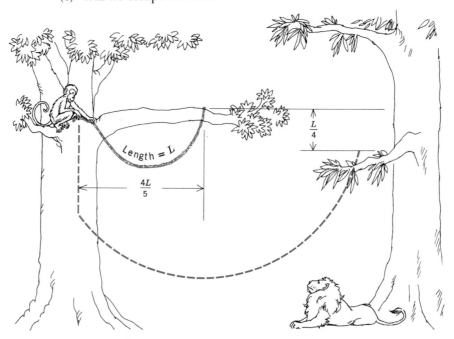

12-32 What choices of the reference point A will result in vanishing of the second term in the right-hand side of Equation 12-13?

12-33 Verify Equation 12-12.

12-34 Using the definition of "moment of absolute momentum" given in the footnote on p. 109, derive the law that relates moment of forces to its rate of change:

$$\mathbf{M}_A = \dot{\mathscr{H}}_A + \mathbf{v}_A \times \mathbf{p}$$

12-35 Referring to the definition of "moment of absolute momentum" given in the footnote on p. 109, show that when the moving point A is taken as the mass center C, the "moment of relative momentum" is equal to the "moment of absolute momentum":

$$\dot{\mathscr{H}}_C = \mathbf{H}_C$$

12-36 Estimate the magnitude of the forces exerted by the bearings A and B on the shaft of a centrifugal pump due to a rock wedged in the impeller as the impeller rotates at ω rad/s. Neglect the effects of the rock on the fluid flow. Indicate the direction and magnitude of the forces on a free-body diagram.

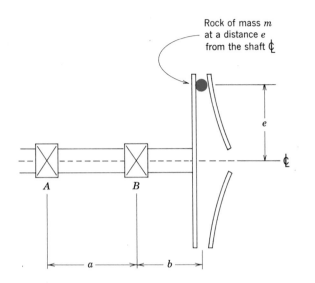

12-37 A water sprinkler consists of a straight tube of length $2L$ and of uniform cross-sectional area A. Water of density ρ enters the tube at the axis of rotation and leaves the tube with a relative velocity v_e. Determine the torque required to rotate the sprinkler at an angular velocity Ω.

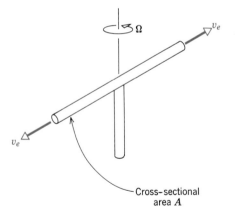

12-38 The rotating machine of Problem 12-14 has had the power shut off, so that the only torque about the rotor axis is that of friction. Assuming this friction torque is proportional to angular velocity, derive the differential equations that describe $y(t)$ and $\phi(t)$.

12-39 Small pendulums such as depicted in the sketch have been used to suppress torsional vibrations in flywheels. In terms of the flywheel angular velocity $\Omega(t)$ and the dimensions shown, write the differential equation that governs $\phi(t)$. Gravity is negligible.

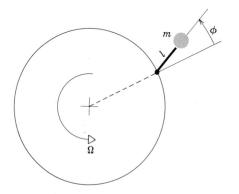

12-40 Show that for small values of ϕ and constant Ω, the motion given by

$$\phi = \phi_0 \cos(\omega_0 t + \theta)$$

will satisfy the differential equation of Problem 12-39. What is the value of the angular frequency ω_0?

12-41 For the motion of the previous problem, what is the moment about the center of the disc of the reaction transmitted to the disc from the pendulum?

12-42 Under some circumstances, the pendulum suspended from the block can suppress vibration induced from the unbalance in the rotor. For constant rotor speed, derive the two differential equations that describe $x(t)$ and $\phi(t)$.

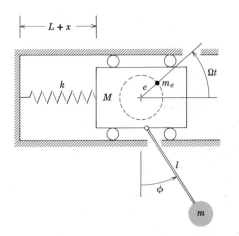

12-43 A body is orbiting around the point O to which the force of attraction is always directed (central force motion).

(a) What does Equation 12-10 imply concerning the angular momentum about the force center?

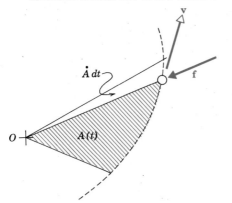

(b) Show that this is equivalent to Kepler's second law of planetary motion (i.e., the rate at which area is swept out by the position vector is constant).

12-44 A satellite orbiting the earth has velocity v_a and distance from the earth's center r_a at apogee. In terms of these, what will be the velocity v_p and distance r_p at perigee? (*Suggestion.* Write down what you know of angular momentum and total mechanical energy.)

12-45 A satellite is launched into orbit around the earth, with initial velocity and altitude as shown. Determine the altitudes and velocities at apogee and perigee. Then determine the length of the semiminor axis of the ellipse.

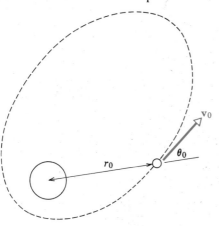

12-46   A 250-kg satellite is launched into space with the velocity and
position shown at cut-off of thrust. Determine the apogee and
perigee of the orbit. What will be the period of the orbit? [*Suggestion*. Recall Kepler's law (Problem 12-43) and the formula for the
area of an ellipse, $A = \pi\, ab$.]

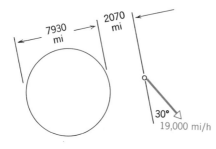

12-47   A particle of mass m is attracted toward a fixed point with a force
that is proportional to the distance from the center, the constant
of proportionality being k. With plane polar coordinates (ρ,ϕ)
oriented in the plane of motion:
(a) Write Newton's second law.
(b) Observe that the expression

$$\rho\, m(\rho\ddot\phi + 2\dot\rho\dot\phi) = \frac{d}{dt}\,(m\rho^2\dot\phi)$$

may be integrated once. What familiar physical law does this
integration lead to?
(c) With initial velocity and radius as shown, what will be the
maximum and minimum radii of the orbit?

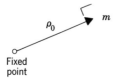

12-48   Two elastically connected objects are moving in a plane in deep
space (removed from gravitational influences). Initially the spring
has a tensile force f_0, and the objects are a distance x_0 apart. The
initial velocities are

$$v_{1\rho} = 6\,v \qquad v_{1\phi} = 8\,v$$
$$v_{2\rho} = 3\,v \qquad v_{2\phi} = 4\,v$$

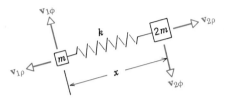

(a) What is the motion of the center of mass?

(b) What are $v_{1\phi}$ and $v_{2\phi}$ when the distance between masses is x?

12-49 A spaceship in a circular polar orbit of 200-mi altitude transfers to a circular orbit of 500 mi by applying a short-time tangential thrust over the South Pole, and again over the North Pole. Sketch the transfer orbit, and evaluate the impulse per unit mass that must be applied over each pole.

12-50 A spaceship in a circular earth orbit inclined 12° to the equator suddenly changes course into a circular equatorial orbit. At what place must the short-time impulse be applied, and what will be its magnitude and direction?

12-51 Two particles are under the influence of a mutual attractive force having magnitude $f(r)$, where r is the distance between the par-

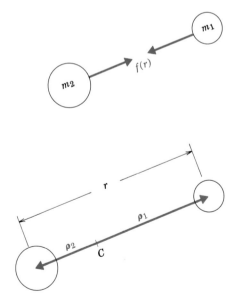

ticles. There are no other forces, so that, from Equation 12-2, the mass center C moves according to

$$(m_1 + m_2)\dot{\mathbf{v}}_C = 0$$

Let $\boldsymbol{\rho}_1$ and $\boldsymbol{\rho}_2$ locate m_1 and m_2 relative to C, so that

$$m_1 \, \boldsymbol{\rho}_1 + m_2 \, \boldsymbol{\rho}_2 = 0$$

(a) Show that

$$\mathbf{H}_C = \mu \, \boldsymbol{\rho}_1 \times m_1 \, \dot{\boldsymbol{\rho}}_1$$

where

$$\mu = 1 + \frac{m_1}{m_2}$$

(b) Considering moments about C, show that

$$\boldsymbol{\rho}_1 \times \dot{\boldsymbol{\rho}}_1 = \mathbf{C} \text{ (constant)}$$

What does this imply concerning the plane of $\boldsymbol{\rho}_1$ and $\dot{\boldsymbol{\rho}}_1$?

(c) Show that Newton's second law applied to m_1 can be written as

$$m_1 \, \ddot{\boldsymbol{\rho}}_1 = -f(\mu \rho_1)\mathbf{u}_1$$

Now compare the above equations with those for a system consisting of a single particle under the influence of a force of attraction toward a fixed point. How much more complex is the two-particle system?

12-52 The mass of the moon is 7.35×10^{22} kg, and the mass of the earth is 5.98×10^{24} kg. The mean distance between the earth and the moon is 3.84×10^5 km. Where is the center of mass of the earth-moon system relative to the earth's surface? Assuming a circular orbit around the mass center, what would be its period? Assuming a circular orbit around the center of the earth, what would be the period?

12-3

WORK AND KINETIC ENERGY. The kinetic energy of a system of particles is defined as the sum of the kinetic energies of the individual particles within the system:

$$T = \sum_i \frac{1}{2} m_i |v_i|^2$$

A form that is often useful expresses this kinetic energy in terms of the total mass and the velocity of a moving reference point A, a term involving the velocities of the point A and the mass center C, and the individual masses and their velocities relative to the point A. (See Problem 12-65).

$$T = \frac{1}{2} m|\mathbf{v}_A|^2 + m\mathbf{v}_A \cdot \mathbf{v}_{C/A} + \frac{1}{2} \sum_i m_i|\mathbf{v}_{i/A}|^2 \qquad (12\text{-}14)$$

The work-kinetic energy relationship, Equation 11-10, of course applies to each particle within the system. If it is written for every particle of the system, and the equations added, we find that the work of all forces is equal to the change in total kinetic energy.

$$\sum_i \int_1^2 \mathbf{f}_i \cdot d\mathbf{r}_i = T_2 - T_1 \qquad (12\text{-}15a)$$

As in the case of the force-acceleration and impulse-momentum laws, the more useful relationships result after the force acting on each particle is separated into the force external to the system and those resulting from other particles within the system.

$$\mathbf{f}_i = \mathbf{f}_{ie} + \sum_j \mathbf{f}_{i/j}$$

However, in the subsequent step of summation over the particles of the system, the *works* of the action-reaction pairs of internal forces do not cancel as did the forces themselves and their impulses. Let us examine this in detail. The sum of the increment of work done on the ith particle by the force from the jth particle and that done on the jth particle by the force from the ith particle we denote by dW_{ij}.

$$dW_{ij} = \mathbf{f}_{i/j} \cdot d\mathbf{r}_i + \mathbf{f}_{j/i} \cdot d\mathbf{r}_j$$

With account of Newton's third law, $\mathbf{f}_{j/i} = -\mathbf{f}_{i/j}$, this becomes

$$\begin{aligned}
dW_{ij} &= \mathbf{f}_{i/j} \cdot d\mathbf{r}_i - \mathbf{f}_{i/j} \cdot d\mathbf{r}_j \\
&= \mathbf{f}_{i/j} \cdot (d\mathbf{r}_i - d\mathbf{r}_j) \\
&= \mathbf{f}_{i/j} \cdot d\mathbf{r}_{i/j}
\end{aligned}$$

Now if the direction of these two forces is that of the line connecting the two particles,

$$\mathbf{f}_{i/j} = -T_{ij} \frac{\mathbf{r}_{i/j}}{r_{i/j}}$$

where T_{ij} is defined as positive when the action-reaction is attractive and negative when it is repulsive. Then,

$$dW_{ij} = -T_{ij} \frac{\mathbf{r}_{i|j} \cdot d\mathbf{r}_{i|j}}{r_{i|j}}$$

$$= -T_{ij} \, dr_{i|j}$$

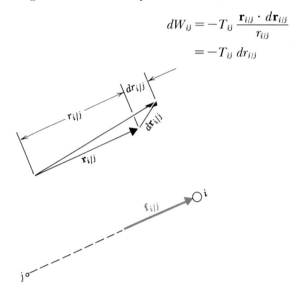

Thus, although the vector sum of the two forces $\mathbf{f}_{i|j}$ and $\mathbf{f}_{j|i}$ is zero, the increment of *work* done on the two particles is proportional to the increment of increase in the distance between the particles.

Now let us introduce this result into the work-kinetic energy relationship (12-15a).

$$\sum_i \int_1^2 \left(\mathbf{f}_{ie} + \sum_j \mathbf{f}_{i|j} \right) \cdot d\mathbf{r}_i = T_2 - T_1$$

$$\sum_i \int_1^2 \mathbf{f}_{ie} \cdot d\mathbf{r}_i - \sum_{i-j} \int_1^2 T_{ij} \, dr_{i|j} = T_2 - T_1$$

Here, the symbol Σ_{i-j} indicates summation over all pairs of particles within the system. [In a system with n particles there are $n(n-1)/2$ such pairs.] Denoting the work of the external forces as W_e, we have

$$\boxed{W_e = \Delta T + \sum_{i-j} \int_1^2 T_{ij} \, dr_{i|j}} \qquad (12\text{-}15\text{b})$$

Thus, in general, part of the work of the external forces results in a change in the total kinetic energy of the system and a part is "absorbed" elsewhere.

As an example, consider the process of the coupling of the two cars (p. 97). The forces external to the two-car system do not work. The change in kinetic energy is

$$\Delta T = \frac{1}{2} (m_1 + m_2) \left(\frac{m_1}{m_1 + m_2} v_0 \right)^2 - \frac{1}{2} m_1 v_0{}^2 = - \frac{1}{2} \frac{m_1 m_2}{m_1 + m_2} v_0{}^2$$

This loss of kinetic energy may be accounted for in the remaining term in Equation 12-15b.

$$\sum_{i-j} \int T_{ij} \, dr_{i/j} = \frac{1}{2} \frac{m_1 m_2}{m_1 + m_2} v_0{}^2$$

The details of the coupling process will be quite complex, and we make no attempt in this two-particle analysis to examine them. We *assume* that some time after contact, all parts of the system are moving at the same speed v_1; the impulse-momentum law then tells us what that speed is, and Equation 12-15b tells us further how much of the initial kinetic energy must have been dissipated.

Conservative Internal Forces. If the magnitude of each internal force is a function only of the corresponding distance between the particles,

$$T_{ij} = T_{ij}(r_{i/j})$$

then the work integral $- \int T_{ij} dr_{i/j}$ is a function only of the distance between the particles (i.e., in no way depends on the manner in which the particles move to reach a particular configuration). This would be the case, for example, with elastic spring interconnections or gravitational interactions. In this case, we can define the potential functions

$$V_{ij}(r_{i/j}) = \int_{(r_{i/j})_0}^{r_{i/j}} T_{ij}(\rho_{i/j}) \, d\rho_{i/j}$$

and, with the sum of these denoted by V:

$$V = \sum_{i-j} V_{ij}$$

the work-energy integral becomes

$$W_e = (T_2 + V_2) - (T_1 + V_1) \tag{12-16}$$

That is, *when the internal forces are all conservative, the work of the external forces is equal to the change in total mechanical energy within the system.*

Example

Referring again to Figure 12-4, suppose the mechanism is "free-wheeling" about the vertical axis (i.e., there is negligible moment transmitted about the vertical axis). Suppose further that friction in the slider is negligible, and that the spring is unstretched when $\chi = 0$. Determine as much as possible about the manner in which χ and $\dot\psi$ vary.

From Equation 12-10, we see that because the moment about the vertical axis is zero, the angular momentum of the two-mass system is conserved:

$$2m(L \sin \chi)^2 \dot\psi = H_0 \text{ (constant)} \tag{a}$$

Also, the only internal action-reaction that can do work on individual particles is that of the spring, and that force is conservative. The external gravitational forces are also conservative, so that the total mechanical energy is conserved. The gravitational potential is

$$V_g = 2mgL(1 - \cos \chi)$$

the potential of the spring is

$$V_{sp} = \frac{1}{2} k [2l(1 - \cos \chi)]^2$$
$$= 2kl^2(1 - \cos \chi)^2$$

and the kinetic energy of the system is

$$T = 2 \left\{ \frac{1}{2} m [(L\dot\chi)_2 + (L \sin \chi \, \dot\psi)^2] \right\}$$
$$= mL^2(\dot\chi^2 + \sin^2 \chi \, \dot\psi^2)$$

The conservation of mechanical energy can thus be written as

$$mL^2(\dot\chi^2 + \sin^2 \chi \, \dot\psi^2) + 2mgL(1 - \cos \chi)$$
$$+ 2kl^2(1 - \cos \chi)^2 = E \text{ (constant)} \tag{b}$$

Now $\dot\psi$ can be eliminated from (a) and (b), and the result arranged in the form

$$\dot\chi^2 + U(\chi) = \frac{E}{mL^2}$$

where

$$U = \frac{2g}{L} (1 - \cos \chi) + \frac{2kl^2}{mL^2} (1 - \cos \chi)^2 + \left(\frac{H_0}{2mL^2} \right)^2 \csc^2 \chi$$

Note that this has the same mathematical structure as Equation 11-3b. Following the procedure there, the variables may be separated, leading to

$$\pm \int \frac{d\chi}{\sqrt{\dfrac{E}{mL^2} - U(\chi)}} = t_0 + t$$

The function defined here is known as an elliptic integral, and is not expressible in terms of the elementary functions with which most readers of this text will be familiar. It predicts oscillatory motion that is approximately sinusoidal for small amplitudes.

In the next chapter we will examine the dynamics of a rigid body, which can be considered as a system in which the distance between every pair of particles remains fixed. What can be said about the last term in Equation 12-15b for a rigid body?

Problems

12-53   Verify that the total mechanical energy is conserved for the example p. 105.

12-54   A satellite circles the earth in a stable circular orbit, traveling at speed v. It is desired to give it a sudden "boost" in speed, just sufficient to cause the satellite to escape the earth's gravitational field. In terms of v, what is this new speed v', which will just permit escape? Note that as r approaches ∞, the velocity should approach zero. Also, after the "boost" $(T + V)$ is constant.

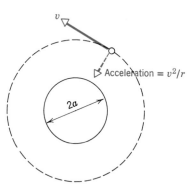

12-55   A 1200-lb lift hoist is traveling along its track at 10 ft/s, carrying a 400-lb load that is swinging, pendulumlike, below. When the load is in the position and at the relative velocity shown, what is the total kinetic energy in the system?

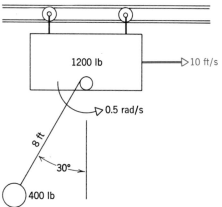

12-56 Compare the kinetic energy of the railroad cars on p. 97 before and after coupling.

12-57 A particle is traveling at speed v_0 as it strikes another particle at rest. The particles have equal mass. During impact, a fraction e of the initial kinetic energy is dissipated. (a) What are the velocities after impact? (b) What is the largest value e can have, and what takes place at this value? (c) What happens as e approaches zero?

12-58 The shuffleboard puck strikes the sitting puck head-on. During impact, 5% of the mechanical energy is dissipated. What are the velocities of the pucks after impact?

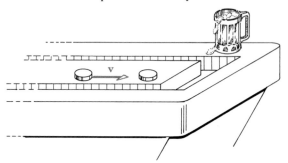

12-59 A 120-Mg engine is coasting at 1.8 m/s when it strikes and becomes coupled with an 18-Mg flatcar that carries a 45-Mg load. The load is not secured to the car, but may slide along the floor, the coefficient of friction being approximately 0.25. (a) Estimate the velocity of the car and engine immediately after coupling. (b) Estimate the velocity after the load has stopped sliding along the flatcar.

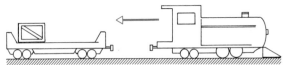

12-60 A chain of length l and mass m is released from rest in the tube as shown in (a). Estimate the velocity of the chain when it reaches the position shown in (b), assuming negligible friction affects.

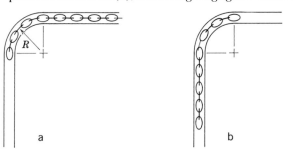

a   b

12-61   Reconsider the chain-and-sprocket drive of Problem 12-27. (a) Evaluate the kinetic energy in the chain. (b) Evaluate the time rate at which the driving torque does work on the system. (c) Neglecting internal friction, use Equation 12-15b to obtain the relationship between torque and angular acceleration of the drive sprocket.

12-62   At a particular instant, the flyball governor of Figure 12-4 is observed to have a motion given by

$$\chi = 37°, \qquad \dot{\chi} = 0, \qquad \dot{\psi} = 2000 \text{ rpm}$$

The mechanism is free of external torque, and its parameters are $L = 150$ mm, $l = 120$ mm, $m = 35$ g, $k = 3.5$ kN/m. Note that initially, since $\dot{\chi} = 0$, χ is at one of its extreme values. Determine the other limit of its oscillation and the value of $\dot{\psi}$ at that extreme. (*Suggestion.* Plot U vs. χ and examine its values at the maximum and minimum χ.)

12-63   Let us reexamine the bullet fired from the sheriff's pistol in Problem 11-60. A reasonable model for the system might be as indicated in the sketch, in which m_1 and m_2 are the masses of the bullet and pistol, respectively. The spring represents the resilience between the pistol and the sheriff's hand. Since the acceleration of the bullet takes place over a very short time, we can reasonably assume that the spring exerts negligible force on the gun until after the bullet has left the barrel. Denote by h the work done by the gas on the bullet and pistol,

$$h = \int f \, dr_{1/2}$$

and by v_1 and v_2 the velocities of the bullet and pistol immediately after firing.

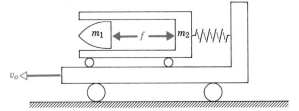

(a) Using (12-5) and (12-15), show that

$$\frac{1}{2} m_1 (v_1 - v_0)^2 = \frac{h}{1 + \dfrac{m_1}{m_2}}$$

(b) If, upon firing from a stationary reference, a 200-grain bullet leaves a 38-oz pistol at 729 ft/s, at what velocity will it leave the chase car?

12-64 Write the equation of conservation of mechanical energy for the system of Problem 12-51, in terms of m_1, ρ_1, $|\dot{\boldsymbol{\rho}}_1|$, and μ.

12-65 Verify Equation 12-14.

12-66 Note that we can integrate Equation 12-2b as follows.

$$\mathbf{f} \cdot \mathbf{v}_C = m\,\dot{\mathbf{v}}_C \cdot \mathbf{v}_C = \frac{d}{dt}\left(\frac{1}{2}\,m\,\mathbf{v}_C \cdot \mathbf{v}_C\right)$$

$$\int_1^2 \mathbf{f} \cdot d\mathbf{r}_C = \frac{1}{2}\,m(v_{C2}{}^2 - v_{C1}{}^2)$$

(a) Is $\int \mathbf{f} \cdot d\mathbf{r}_C$ the work of the external forces on the system?
(b) Is $(1/2)m\,v_C{}^2$ the kinetic energy of the system?
(c) Can we say that the work of the external forces will equal the change in kinetic energy of the system? Explain.

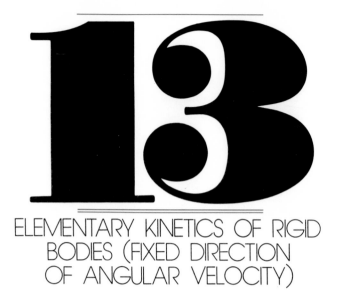

ELEMENTARY KINETICS OF RIGID BODIES (FIXED DIRECTION OF ANGULAR VELOCITY)

The dynamics of rigid bodies presents many interesting and surprising phenomena. For an illustration, the following experiment is suggested. With a stiff rubber band, secure this or some other book closed, and toss it into the air, spinning in three different ways. First, spin it about an axis perpendicular to the covers; next, about an axis through the pages from top to bottom and, finally, about an axis through the pages from left to right. Note that the direction of the angular velocity vector does not always remain fixed, even though in flight there is negligible moment about the mass center! An explanation of your observations will be undertaken in Chapter 16. There we will see that if the mass is distributed somewhat like that of a plate that is relatively thin and spinning about an axis perpendicular to the plane of the plate, the direction of the angular velocity tends to remain constant.

In this chapter we will *assume* that the direction of the angular velocity remains fixed. But we must remember that meeting this condition may require that the body be mechanically constrained (say, by mounting it in bearings with a shaft).

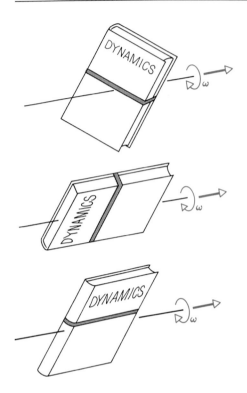

A rigid body may be treated as a special case of a system of particles, so that the laws expressed by Equations 12-2 and 12-13 apply here as well.

$$\mathbf{f} = m\,\mathbf{a}_C \qquad\qquad [12\text{-}2]$$

$$\mathbf{M}_A = \dot{\mathbf{H}}_A + \mathbf{r}_{C/A} \times m\mathbf{a}_A \qquad [12\text{-}13]$$

Application of the first of these equations requires no new considerations here. Expressions for the angular momentum and its rate of change will require further development before the second equation is ready for application to rigid bodies.

13–1

ANGULAR MOMENTUM.   The definition of angular momentum is as for the system of n particles (Equation 12-11) but, since we are now considering a continuous distribution of mass, the summation is carried out as an integration over the body.

$$\mathbf{H}_A = \int_m \mathbf{r}_{P/A} \times \mathbf{v}_{P/A}\,dm$$

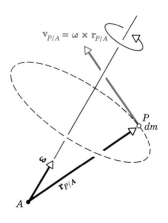

Here P indicates the point at which the element of mass dm is located. If we let point A be attached to the body, the velocity difference is completely determined by the angular velocity through Equation 10-11. The angular momentum can then be written directly in terms of $\boldsymbol{\omega}$ by inserting the relationship

$$\mathbf{v}_{P/A} = \boldsymbol{\omega} \times \mathbf{r}_{P/A}$$

into the above integral. Without carrying the subscripts through the remainder of the development, it is to be understood that the point A is the reference for position $\mathbf{r}$, entering the definition of angular momentum.

$$\mathbf{H} = \int_m \mathbf{r} \times (\boldsymbol{\omega} \times \mathbf{r}) \, dm$$

The vector identity (3-13) may be used to rewrite this as

$$\mathbf{H} = \boldsymbol{\omega} \int_m \mathbf{r} \cdot \mathbf{r} \, dm - \int_m (\boldsymbol{\omega} \cdot \mathbf{r})\mathbf{r} \, dm$$

Now let us introduce a set of rectangular Cartesian coordinate axes emanating from the reference point A, with the z axis coinciding with the direction of the angular velocity. Then,

$$\boldsymbol{\omega} = \omega \, \mathbf{u}_z$$

$$\boldsymbol{\omega} \cdot \mathbf{r} = \omega z$$

and the angular momentum vector can be expressed as

$$\mathbf{H} = \omega \mathbf{u}_z \int_m (x^2 + y^2 + z^2) \, dm - \omega \int_m z(x\mathbf{u}_x + y\mathbf{u}_y + z\mathbf{u}_z) \, dm$$

The component of angular momentum in the direction of $\boldsymbol{\omega}$ is

$$H_z = \mathbf{H} \cdot \mathbf{u}_z$$

$$= \omega \int_m (x^2 + y^2 + z^2)\, dm - \omega \int_m z^2\, dm$$

$$\boxed{H_z = I_{zz}\, \omega} \qquad (13\text{-}1a)$$

where

$$\boxed{I_{zz} = \int_m (x^2 + y^2)\, dm} \qquad (13\text{-}2a)$$

The components of angular momentum in the directions perpendicular to $\boldsymbol{\omega}$ are

$$H_x = \mathbf{H} \cdot \mathbf{u}_x \qquad\qquad H_y = \mathbf{H} \cdot \mathbf{u}_y$$

$$\boxed{H_x = I_{xz}\omega} \qquad\qquad \boxed{H_y = I_{yz}\omega} \qquad (13\text{-}1b)$$

where

$$\boxed{I_{xz} = -\int_m xz\, dm} \qquad\qquad \boxed{I_{yz} = -\int_m yz\, dm} \qquad (13\text{-}2b)$$

Products of Inertia. The integrals defined by Equations 13-2 are called *second moments of mass*, or *products of inertia* of the body; I_{zz} is usually referred to as the *moment of inertia* about the z axis. They represent properties of the body that depend on the distribution of mass throughout the body, much the same way that the second moments of area defined in Equations 6-7 depend on the size and shape of the plane area. Values for several homogeneous bodies are tabulated in Appendix C.

As in the case of second moments of plane areas, it is helpful to have formulas to relate the products of inertia with respect to a given set of axes to those with respect to other sets of axes. The relationships for two sets of parallel axes can be determined in the same way that led to the analogous formulas for second moments of area (Equations 6-8, Section 6-2). If the axes X, Y, Z emanating from the mass center C are parallel to the axes x, y, z emanating from an arbitrary point A, the products of inertia with respect to the two sets of axes are related by

$$I_{zz} = I_{ZZ} + m(c_x^2 + c_y^2)$$
$$I_{xz} = I_{XZ} - m\, c_x\, c_z \qquad (13\text{-}3)$$
$$I_{yz} = I_{YZ} - m\, c_y\, c_z$$

where c_x, c_y, and c_z are the x, y, z coordinates of the center of mass C. Thus each of the products of inertia may be computed by adding, to the corresponding product of inertia with respect to parallel axes through the mass center, a term representing the product of inertia the body would have if the mass were concentrated at C.

Example

A rigid body consists of two particles supported by a plane truss of negligible mass. It is spinning in rigidly supported bearings, as shown in Figure 13-1a. Compute the products of inertia relative to the coordinate

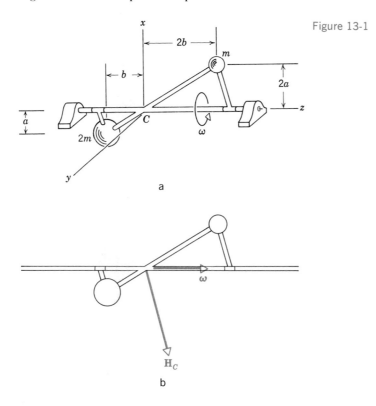

Figure 13-1

a

b

axes shown, evaluate the angular momentum about the mass center, and show it on a sketch of the body.

The products of inertia are

$$I_{zz} = (2a)^2(m) + a^2(2m) = 6ma^2$$
$$I_{xz} = -[(2a)(2b)(m) + (-a)(-b)(2m)] = -6mab$$
$$I_{yz} = -(2b)(0)(m) + (-b)(0)(m) = 0$$

Using these values in Equations 13-1, we find

$$\mathbf{H} = 6ma^2\omega\ \mathbf{u}_z - 6mab\omega\ \mathbf{u}_x$$
$$= 6ma\omega(a\mathbf{u}_z - b\mathbf{u}_x)$$

This vector is shown in Figure 13-1b.

Problems

13-1   Compute I_{zz} and I_{zx} for the thin rod. Evaluate and sketch the angular momentum vector for a spinning about the z axis.

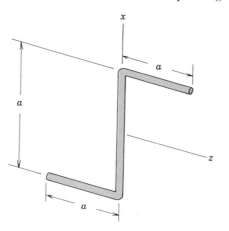

13-2   Compute I_{xx}, I_{zz}, I_{yy}, and I_{zx} of the thin, homogeneous, rectangular plate.

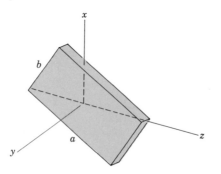

13-3   The plate of Problem 13-2 is mounted in rigidly supported bearings and spinning uniformly about the z axis. Compute the angular momentum and show this vector on a sketch. Compute the moment of forces about the center of the plate, acting on the plate, and the reactions from the bearings. Draw a free-body diagram to show directions.

13-4 Suppose the "particles" of the example on p. 133 are actually uniform steel spheres of diameters 20 mm and 25.2 mm, respectively, $a = 60$ mm and $b = 90$ mm. Still neglecting the mass of the support truss, evaluate I_{zz} and I_{zx}. How do these values compare with those computed on the basis that the balls are particles?

13-5 Determine the indicated second moments of mass of the following bodies by direct integration and by means of the information in Appendix C

(a) Sphere

$I_{zz} = ?$

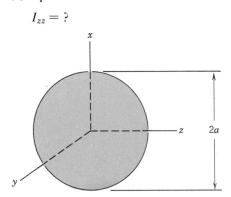

(b) Right circular cone

$I_{xx} = ?$
$I_{zz} = ?$
$I_{\xi\xi} = ?$

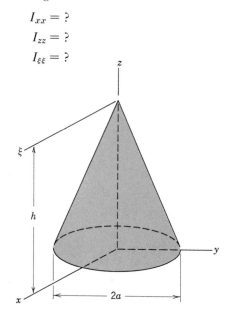

(c) Thin circular plate

$$I_{xx} = ?$$
$$I_{zz} = ?$$

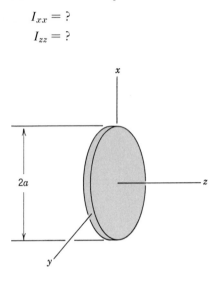

(d) Slender circular
hoop

$$I_{xx} = ?$$

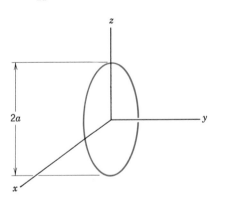

13-6 Verify the formulas for second moments of mass for the thin flat plate (p. 176).

13-7 Verify the formulas for second moments of inertia of the cylinder (p. 176).

13-8 Verify the formulas for second moments of mass of the cone (p. 177).

13–2

ROTOR UNBALANCE. Observe that for a given orientation of $\boldsymbol{\omega}$ in the rotating body, the corresponding **H** is determined by the manner in which mass is distributed with respect to the rotation axis. Thus, if $\boldsymbol{\omega}$ remains fixed with respect to the body, so will **H**. Now, if the body in Figure 13-1a spins at constant rate, the head of the vector **H** will move in the circular path shown in Figure 13-2, and $\dot{\mathbf{H}}$ will be tangent to this circle. In

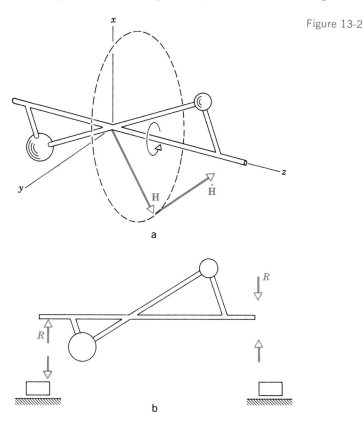

Figure 13-2

a

b

fact, the situation here is exactly like that indicated in Figure 10-10 for the vector **r**, the rate of change of which was shown to be $\dot{\mathbf{r}} = \boldsymbol{\omega} \times \mathbf{r}$ (Equation 10-10). Thus we have, for constant $\boldsymbol{\omega}$,

$$\dot{\mathbf{H}} = \boldsymbol{\omega} \times \mathbf{H}$$

In terms of the coordinate directions in Figure 13, this has the value

$$\dot{\mathbf{H}} = \omega H_x \mathbf{u}_y$$
$$= -I_{zx} \, \omega^2 \mathbf{u}_y \tag{13-4}$$

But, from Equation 12-13, this is equal to the moment about C of forces applied to the body. These forces must be the reactions from the bearings, as shown in the free-body diagram in Figure 13-2b.

This example typifies the so-called *dynamically unbalanced rotor*. Whenever the product of inertia I_{zx} or I_{zy} is not zero, there exists a rotating moment component perpendicular to the axis of rotation, z. Since the bearing reactions that constitute this moment are normally undesirable, such a spinning rotor is designed so that the products of inertia I_{zx} and I_{zy} are as small as possible.

At high rates of rotation, such as in a turbine, a small unbalance of this type can induce severe vibration difficulties. After final assembly in manufacture, a rotor may be spun in a balancing machine that detects the amount and orientation of unbalance. Small corrective mass may then be added to reduce the objectionable products of inertia.

Example

A 270-kg rotor is spun at 840 rad/s. Rotating reactive forces of 370 N, acting in exactly opposite directions, are detected at each of two bearings placed 1.7 m apart. Provision is made at positions 0.9 m apart along the axis to add corrective mass. Recommend amounts and locations of corrective mass.

According to Equation 12-2, if the center of mass were not on the spin axis, the centripetal acceleration associated with its circular motion would induce bearing reactions having a resultant not equal to zero. But, since the given reactions form a pure couple, the center of mass must lie on the spin axis. So that the corrective mass does not shift the mass center away from the axis, the mass center of the added material must also lie on the spin axis.

The moment of the couple has the magnitude

$$M = (370 \text{ N})(1.7 \text{ m})$$
$$= 629 \text{ N·m}$$

But, from Equations 12-13 and 13-4,

$$I_{zx} = \frac{M}{\omega^2}$$

$$= \frac{(629 \text{ N·m})}{(840 \text{ rad/s})^2}$$

$$= 8.91 \times 10^{-4} \text{ kg·m}^2$$

Referring to Figure 13-3, we see that the product of inertia of the corrective mass will be

$$I'_{zx} = -[m_1\rho_1 z_1 + m_2(-\rho_2)(z_1 - 0.9 \text{ m})]$$

Figure 13-3

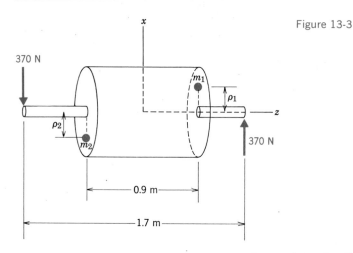

370 N

x

m_1

ρ_1

z

ρ_2

m_2

370 N

0.9 m

1.7 m

But, because the mass center must remain on the axis of rotation,

$$m_2\rho_2 = m_1\rho_1$$

so that

$$I'_{zx} = -m_1\rho_1[z_1 - (z_1 - 0.9 \text{ m})] = -m_1\rho_1(0.9 \text{ m})$$

Setting $I_{zx} + I'_{zx} = 0$ yields

$$m_1\rho_1 = \frac{8.91 \times 10^{-4} \text{ kg·m}^2}{0.9 \text{ m}}$$
$$= 0.990 \text{ g·m}$$

Addition of 10 g at a radius of 99 mm at each location will do the job.

Problems

13-9 The circular disk of mass m and radius a is mounted eccentrically on the shaft as shown, and the assembly is rotating at N rpm. Estimate the bearing forces on the shaft due to the eccentricity when in the position shown. Indicate both directions and magnitudes on a free-body diagram.

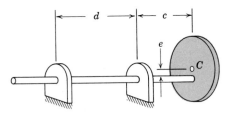

d

c

e

C

13-10 Tests indicate that when turning at 1800 rpm the statically balanced motor armature exerts a rotating moment on the bearings of 5000 ft·lb. How deep should properly oriented 1 in. diameter, longitudinal holes, located as shown, be drilled to dynamically balance the rotor? The material removed by drilling has a mass of 0.2 lbm per inch of hole depth.

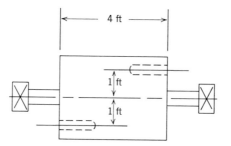

13-11 The two throw, opposed crank crankshaft shown in the schematic is to be dynamically balanced by the addition of mass on 3-in arms at A and B. Each crank can be considered as providing the equivalent of a 6-lbm eccentric mass located at the center of the throw and 2 in. off the crank shaft axis of rotation. Estimate the amount of additional mass required and indicate their locations on a sketch.

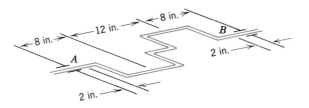

13-12 The rotating bearing reactions shown in the sketch are measured during a spin test of the 80-kg rotor at 1200 rpm.
(a) Determine the location of the mass center.
(b) Indicate amounts and locations in planes A and/or B of added mass necessary to move the center of mass to the spin axis.
(c) This corrective mass having been added, the reactions may be expected to be of equal magnitude and opposite direction. Determine locations and amounts of additional mass in planes A and B that will complete the balancing.

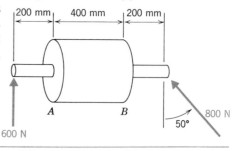

13–3

EQUATIONS OF TWO-DIMENSIONAL MOTION.   We turn now to the component of moment about an axis in the direction of $\boldsymbol{\omega}$. From Equations 12-13 and 13-1, we have

$$M_z = \mathbf{M}_A \cdot \mathbf{u}_z$$
$$= \dot{\mathbf{H}}_A \cdot \mathbf{u}_z + (\mathbf{r}_{C/A} \times m\mathbf{a}_A) \cdot \mathbf{u}_z$$

$$\boxed{M_z = I_{zz}\dot{\omega} + m(\mathbf{r}_{C/A} \times \mathbf{a}_A) \cdot \mathbf{u}_z} \qquad (13\text{-}5)$$

Whenever the z-axis passes through the mass center, or the acceleration of point A is zero or in the direction of either $\mathbf{r}_{C/A}$ or $\mathbf{u}_z$, the last term in Equation 13-5 vanishes, and the moment about the z axis is related to the angular acceleration ω through

$$M_z = I_{zz}\dot{\omega} \qquad (13\text{-}5a)$$

Example

The homogeneous circular cylinder rolls without slipping on the circular cylindrical surface, as shown in Figure 13-4. Determine the equation governing its motion in terms of the angle $\phi(t)$.

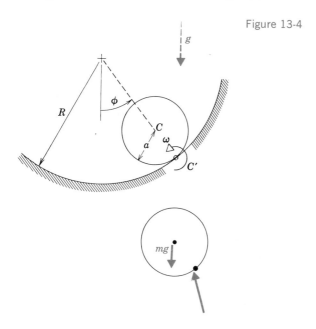

Figure 13-4

The angular speed ω of the cylinder may be determined in terms of $\dot{\phi}$ by considering the velocity of point C from two points of view; that is, in terms of the rotation rate of the line connecting the center of the circular surface and the center of the rolling cylinder, and in terms of the rotation of the cylinder about the instantaneous axis C'.

$$v_C = (R - a)\dot{\phi} = -a\omega$$

where the positive direction for ω is defined as counterclockwise. This yields

$$\omega = -\frac{R - a}{a}\dot{\phi} \tag{a}$$

The acceleration of C' can be shown to be directed toward the mass center C, so that (13-5a) applies. Referring to Appendix C and (13-3), we find the moment of inertia about the instantaneous axis to be

$$I_{C'} = \frac{ma^2}{2} + ma^2 = \frac{3ma^2}{2} \tag{b}$$

From the free-body diagram of the cylinder, (Figure 13-4b), the moment of forces about the instantaneous axis is

$$M_{C'} = mga \sin \phi \tag{c}$$

Substitution of (a), (b), and (c) into (13-5a) yields

$$\ddot{\phi} + \frac{2g}{3(R - a)} \sin \phi = 0$$

If the motion is restricted to small amplitudes, the approximation $\sin \phi \approx \phi$ is valid, reducing the equation of motion to

$$\ddot{\phi} + \frac{2g}{3(R - a)} \phi = 0$$

An equation of the same mathematical form was solved earlier, as an example application of Equations 11-3. This equation,

$$\ddot{x} + \frac{k}{m} x = 0$$

governing the rectilinear motion of the spring-suspended mass, has the solution

$$x = x_0 \cos \left(\sqrt{\frac{k}{m}} \, t + \theta \right)$$

Therefore the solution for the equation governing small ϕ is

$$\phi = \phi_0 \cos \left[\sqrt{\frac{2g}{3(R - a)}} \, t + \theta \right]$$

It is instructive to repeat the derivation of the equation of motion by considering moments of force and momentum about the axis through the center of mass.

Radius of Gyration and Center of Percussion.   When the direction of **ω** is fixed, it is sometimes convenient to view an "equivalent" body—equivalent in the sense that its mass and moment of inertia I_{zz} are the same as those of the actual body, but in which the mass is all concentrated at the same distance from the z axis. This distance is called the *radius of gyration,* k, defined by

$$mk^2 = I_{zz}$$ (13-6)

For the two-dimensional case, a related distance from the reference point A is that to the *center of percussion, P_A.* Consider a force that acts for a short duration with its line of action passing perpendicularly through the line connecting the reference point A and the center of mass. If the intersection is at the center of percussion, this "impulsive" force will produce

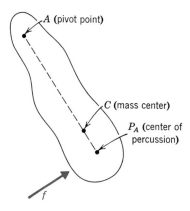

no change in the velocity of the reference point A. In other words, if the body is at rest and unsupported, the initial effect of an impulsive force at P_A is to cause the body to rotate about A. (A baseball striking this point on a bat will transfer a minimum "shock" to the batter's hands.) To determine the location of P_A, we first integrate Equation 13-5 with respect to time.

$$\int \overline{AP}_A f \, dt = I_A \, \Delta\omega + m\mathbf{u}_z \cdot \int \mathbf{r}_{C/A} \times d\mathbf{v}_A$$

But, since the impulse is to produce no change in the velocity of A,

$dv_A = 0$. The left-hand member can be evaluated with the help of the linear impulse-momentum relationship

$$\int \overline{AP}_A f \, dt = \overline{AP}_A \int f \, dt = \overline{AP}_A m \, \Delta v_C$$

Substitution of this and the kinematic relationship

$$\Delta v_C = \overline{AC} \, \Delta \omega$$

into the above yields

$$\overline{AC} \; \overline{AP}_A = \frac{I_A}{m} = k^2 \tag{13-7}$$

Thus the radius of gyration is the root-mean-square of the distances from A to the center of mass and to the center of percussion. The distance to the center of percussion can also be interpreted as an "equivalent" length of a pendulum pivoted about point A. (See Problem 13-39.)

Observe that, unlike the center of mass, the center of percussion is not a point inherent solely to the body itself; its location depends on the choice of the reference point A as well as the distribution of mass in the body.

Example

To determine its moment of inertia, a 40-oz rigid body is suspended from a fixed pivot A, as shown in Figure 13-5. The period of small oscillation is

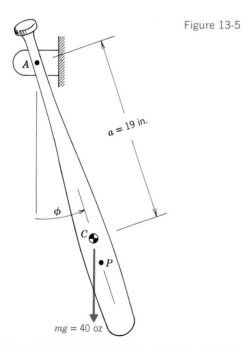

Figure 13-5

measured to be 1.49 s. By balancing the body, the mass center has been determined to lie 19 in. from the pivot point. Determine the radius of gyration with respect to the point A and with respect to the mass center.

An application of Equation 13-5a leads to

$$I_A \ddot{\phi} + mga \sin \phi = 0$$

The simplified form of this equation,

$$\ddot{\phi} + \frac{mga}{I_A} \phi = 0$$

valid for small oscillations, has appeared in a number of earlier examples. Its solution exhibits oscillations at the angular frequency

$$\omega_0 = \sqrt{\frac{mga}{I_A}}$$

Therefore the period of oscillation is

$$p = \frac{2\pi \text{ rad/cycle}}{\omega_0 \text{ rad/unit time}}$$

$$= 2\pi \sqrt{\frac{I_A}{mga}} \text{ time units/cycle}$$

Solving for the radius of gyration, we obtain

$$k_A = \sqrt{\frac{I_A}{m}}$$

$$= \frac{p \sqrt{ga}}{2\pi}$$

$$= \frac{(1.49 \text{ s}) \sqrt{(386 \text{ in/s}^2)(19 \text{ in.})}}{2\pi}$$

$$= 20.3 \text{ in.}$$

We note in passing that the center of percussion is at a distance

$$AP_A = \frac{I_A}{ma} = \frac{k_A^2}{a} = 21.7 \text{ in.}$$

from the pivot point.

The radius of gyration with respect to the mass center can be determined using Equation 13-3:

$$I_A = I_C + ma^2$$

which gives

$$k_C = \sqrt{k_A^2 - a^2}$$
$$= 7.17 \text{ in.}$$

Problems

13-13   Estimate the force exerted by workman A on the plank of Problem 8-8.

13-14   The sphere rolls down the incline, without slipping, under the influence of gravity. Write the differential equation that governs $x(t)$.

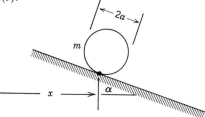

13-15   The center of the cylinder is given an initial speed v_0 up the incline, and the initial angular velocity is ω_0 in the direction shown. The coefficient of friction is μ.

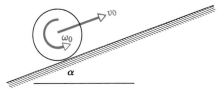

(a) Determine how high the cylinder travels before slipping stops.
(b) What maximum height will the cylinder reach?

13-16   Sprocket A has a moment of inertia I_A about its axis of rotation and sprocket B a moment of inertia I_B about its axis of rotation. The driving chain has total mass m. What angular acceleration of

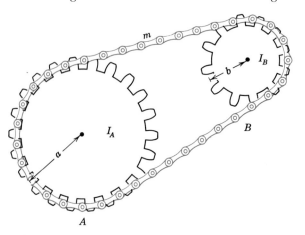

sprocket B will be produced by a torque T applied to sprocket A? What angular acceleration of sprocket A will be produced by a torque T applied to sprocket B?

13-17   Estimate the values of F_1 and F_2 when the brakes are locked and the 1.25-ton car is sliding in a straight line. Take the coefficient of friction between the wheels and the pavement to be approximately 0.80.

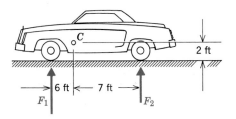

13-18   The ship is dead in the water when the tug exerts a force on it in the direction shown of 3000 lb. The ship has a moment of inertia of 8×10^{10} lbf·ft·s² about a vertical axis through its center of mass and has a displacement (mass) of 10 000 tons. Estimate the location of the point about which the ship will tend to rotate. Assume that until the ship acquires a substantial angular and translatory velocity, the water offers negligible resistance to its movement.

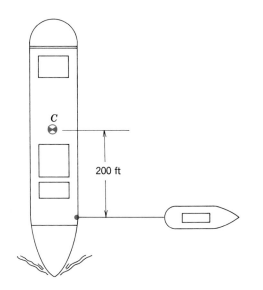

13-19   The beam of mass m is supported by a pin connection at each end. The pin in the left end suddenly fails completely, providing no support for the left end. Determine the change in the value of the reaction at the right end.

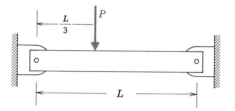

13-20   The two different objects are suspended from the pulley that has moment of inertia I and is free to turn with negligible friction. What will be the accelerations of the objects?

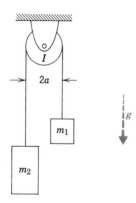

13-21   The soapbox racer and driver have a total mass M. Each wheel has mass $m/4$, radius a, and radius of gyration k. How long will it take the racer to travel a distance s, starting from rest? Would there be an advantage to be gained by adding mass to the car? What would be the primary consideration in selecting the wheels for the racer?

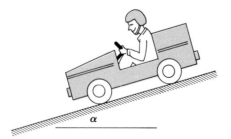

13-22 Work the example problem on pp. 141–142 using the mass center *C* as reference for moment and angular momentum.

13-23 What is the smallest coefficient of friction that will prevent slipping of the roller in the example on pp. 141–142?

13-24 The homogeneous semicircular cylinder rolls without slipping on the horizontal surface. Derive the differential equation governing the angle $\theta(t)$ in terms of quantities shown on the sketch. By comparison with the other systems that have led to the same differential equation, determine the frequency of small oscillations.

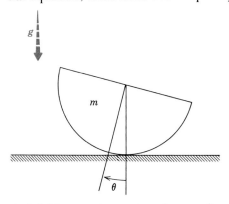

13-25 A 1.5-Mg car is accelerating at 2 m/s². Each drive wheel has a mass of 16 kg, an outside diameter of 610 mm, and a radius of gyration of 230 mm. Draw free-body diagrams of the car, the car minus wheels, and the wheels. Evaluate the torque applied to the drive wheels.

13-26 An airplane of mass *m* is making a rough landing, with maximum compressive force in the landing gear of f_{max}. The moment of inertia of the airplane, about an axis perpendicular to the paper and through point *A*, is I_A. What is the maximum angular acceleration of the airplane?

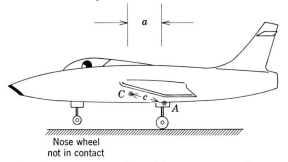

Nose wheel
not in contact

Check your answer by taking moments about a second point.

13-27 A 10-in., 16-lb ball is dropped with a mass center velocity of 40 ft/s and an angular velocity of 2 rad/s on a polished maple floor. Assuming a coefficient of friction of 0.25, estimate the time required for the ball to stop sliding (i.e., begin rolling without sliding), and its mass center velocity at that time. (Rotation is in plane of motion.)

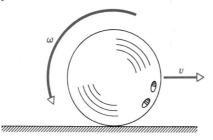

13-28 There is sufficient friction to prevent slipping at the surface where the wheel makes contact. Derive the differential equation that governs $x(t)$.

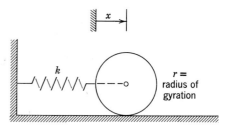

13-29 What is the minimum coefficient of friction that will prevent slipping of the cylinder in the previous problem?

13-30 The two monkeys, each having completed a novel, are now experimenting in mechanics. They are of equal mass and balance, stationary, on either side of the pulley. The mass of the rope and the friction in the pulley bearing are negligible. Socrates begins to climb. What happens to Plato? Explain in terms of fundamental laws of mechanics.

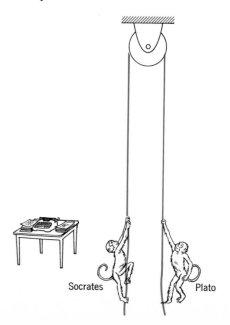

Socrates   Plato

13-31   The end of the yo-yo string is being accelerated upward at the rate a_0. The mass of the yo-yo is m, its radius of gyration is k, and the spool diameter is d. Determine the tension in the string.

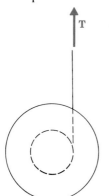

13-32   Consider an airplane fitted with an auxiliary 10-gal fuel tank on each wing-tip (wing-span is 24 ft). What will be the additional moment required to produce a given angular deceleration rate (0.1 rad/s²) from a spin angular velocity about the z axis of 0.3 rad/s when the tip tanks are full, over that required when they empty? The density of the fuel is approximately 6 lbm/gal.

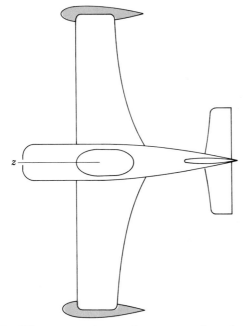

13-33   The curve shows the torque that the field of an electric motor applies to the armature. In terms of T_0, ω_1, and the moment of inertia of the armature I, estimate the time required for the motor to accelerate from rest to its equilibrium speed with no external load. (*Suggestion.* Note that we have here the rotational analogue of the class of force discussed on pp. 64–65.)

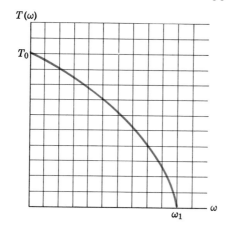

13-34 To determine experimentally the moment of inertia of a large ship's propeller, it is hung as sketched on a small steel bar and its natural frequency of torsional vibration measured. Also measured is the constant α in the linear torque-twist relationship for the bar.
  (a) Derive the differential equation that describes the rotational displacement of the propeller.
  (b) Develop the relationship that gives the moment of inertia of the propeller in terms of the stiffness α and the period of oscillation T.

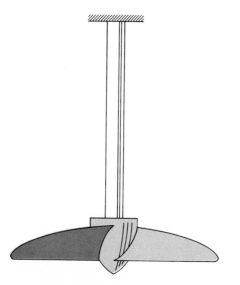

13-35 The radius of gyration of an automobile wheel is to be determined by measuring its oscillation in the support shown. Develop the relationship that gives the moment of inertia about the axis through the center of mass in terms of the period of oscillation and the quantities indicated on the sketch.

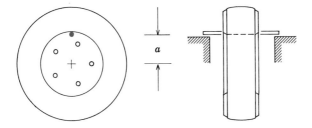

13-36 The pendulum of a Charpy impact testing device has mass m. The center of mass has been located by balancing at a distance d from its axis of rotation. When allowed to swing freely back and forth about its axis of rotation, it is observed to have a period of T. Estimate the moment of inertia of the pendulum about its axis of rotation.

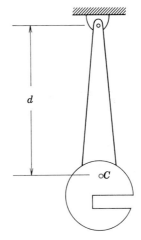

13-37 In an experiment to determine the moment of inertia of a wheel about its axis of rotation, it is mounted as shown on a slender rod having a torsional stiffness of 19 lbf·in/rad. The frequency of vibration of the 30-lb wheel is found to be such that 10 cycles are completed in 12.0 s. Estimate the radius of gyration.

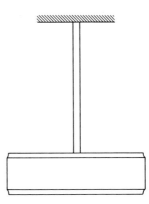

13-38 A rotor assembly of mass m and shaft diameter D is supported by the shaft on two curved tracks and set to rolling back and forth without slipping. The natural frequency of oscillation for small displacements is f Hz. Determine the moment of inertia of the rotor assembly about an axis of rotation through the center of mass.

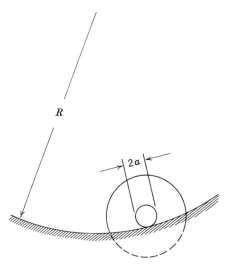

13-39 The pendulum consists of a rigid body pivoted at the support at A. The moment of inertia about an axis perpendicular to the paper through A is I_A. Write the differential equation governing the angle $\theta(t)$. Compare with the equation governing the pendulum idealized as a particle on a stiff, massless rod. What is the "equivalent" length of this pendulum?

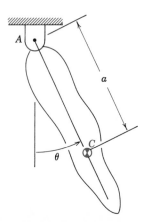

13-40 Show that the radius of gyration is always greater than the distance from the reference axis to the center of mass, and that consequently the distance to the center of percussion is still greater. What happens to the distance to the center of percussion as we let the reference point A approach the center of mass M?

13-41 As the bat meets the baseball, its motion may be resolved as a rotation about an instantaneous axis located somewhere along the batter's arms. In order to transmit minimum "shock" to the batter's hands, the ball should strike the center of percussion with reference to what point?

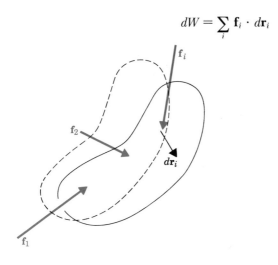

13-4

WORK AND KINETIC ENERGY FOR A RIGID BODY.   The increment of work done by the forces $\mathbf{f}_1$, $\mathbf{f}_2$, . . . , $\mathbf{f}_n$ as the points of application on the body undergo increments of displacement $d\mathbf{r}_1$, $d\mathbf{r}_2$, . . . , $d\mathbf{r}_n$ is defined as

$$dW = \sum_i \mathbf{f}_i \cdot d\mathbf{r}_i$$

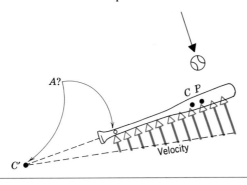

This can also be expressed in terms of the shift in position of an arbitrary point A attached to the body and an increment of rotation about an axis through this point. For this purpose, let $\mathbf{r}_{i/A}$ locate the point of application of $\mathbf{f}_i$ relative to the point A.

$$\mathbf{r}_i = \mathbf{r}_A + \mathbf{r}_{i/A}$$

Then,

$$dW = \sum_i \mathbf{f}_i \cdot (d\mathbf{r}_A + d\mathbf{r}_{i/A})$$

$$= \sum_i \mathbf{f}_i \cdot \mathbf{v}_A \, dt + \sum_i \mathbf{f}_i \cdot \dot{\mathbf{r}}_{i/A} \, dt$$

$$= \left(\sum_i \mathbf{f}_i \right) \cdot \mathbf{v}_A \, dt + \sum_i \mathbf{f}_i \cdot (\boldsymbol{\omega} \times \mathbf{r}_{i/A}) \, dt$$

$$= \mathbf{f} \cdot \mathbf{v}_A \, dt + \sum_i (\mathbf{r}_{i/A} \times \mathbf{f}_i) \cdot \boldsymbol{\omega} \, dt$$

$$= \mathbf{f} \cdot \mathbf{v}_A \, dt + \mathbf{M}_A \cdot \boldsymbol{\omega} \, dt \tag{13-8}$$

The first term is exactly like the increment of work for a particle: the product of the component of force in the direction of the increment of displacement of the particle, and the magnitude of this increment of displacement. The additional part is the product of the component of the resultant moment in the direction of the increment of rotation about point A and the magnitude of this increment of rotation.

Equation 12-14 gives the kinetic energy of a system of particles in terms of the velocity of a reference point A and velocities of particles relative to A; for a body with distributed mass the summation is carried out as an integration over the body.

$$T = \frac{1}{2} m v_A^2 + m \mathbf{v}_A \cdot \mathbf{v}_{C/A} + \frac{1}{2} \int_m v_{P/A}^2 \, dm \tag{13-9a}$$

Here P is the point at which the element of mass dm is located. Now, if the body is simultaneously translating with velocity $\mathbf{v}_A$ and rotating about an axis through A with angular velocity $\boldsymbol{\omega}$, we have, from Equation 10-11,

$$\mathbf{v}_{C/A} = \boldsymbol{\omega} \times \mathbf{r}_{C/A}$$
$$v_{P/A} = |\boldsymbol{\omega} \times \mathbf{r}_{P/A}|$$
$$= \rho \omega$$

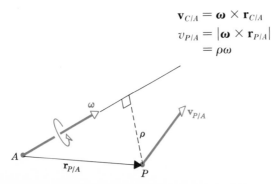

where ρ is the perpendicular distance, from the axis through A in the direction of $\boldsymbol{\omega}$, to point P. With the use of these relationships, the kinetic energy may be expressed in terms of the velocity of A and the angular velocity, as

$$T = \frac{1}{2} m v_A{}^2 + m \mathbf{v}_A \cdot (\boldsymbol{\omega} \times \mathbf{r}_{C/A}) + \frac{1}{2} I_{A\omega} \, \omega^2 \qquad \text{(13-9b)}$$

where

$$I_{A\omega} = \int_m \rho^2 \, dm$$

is the moment of inertia about the axis through A in the direction of $\boldsymbol{\omega}$. If the reference point A is chosen at the center of mass, this reduces to

$$T = \frac{1}{2} m v_C{}^2 + \frac{1}{2} I_{C\omega} \, \omega^2 \qquad \text{(13-9c)}$$

Work-Kinetic Energy Integral for Fixed Direction of Angular Velocity.
Integration of the equation of rotational motion proceeds in the same manner as for the translational motion of a particle. We begin by choosing the reference point A at the center of mass and multiplying each member of Equation 13-5 by an increment of angular displacement:

$$M_{Cz} \, d\theta = I_{C\omega} \frac{d\omega}{dt} \, d\theta$$

$$= I_{C\omega} \frac{d\theta}{dt} \, d\omega$$

$$= d \left(\frac{1}{2} I_{C\omega} \omega^2 \right)$$

Integration gives us the relationship between the work associated with rotation about the center of mass and the kinetic energy associated with the corresponding motion.

$$\int_1^2 M_{Cz} \, d\theta = \left[\frac{1}{2} I_{C\omega} \omega^2 \right]_1^2 \qquad \text{(13-10)}$$

Note the analogy between this expression and that involving the translation of the center of mass.

$$\int_1^2 \mathbf{f} \cdot d\mathbf{r}_C = \left[\frac{1}{2} m v_C{}^2 \right]_1^2 \qquad \text{(13-11)}$$

The subscript C is pivotal in both Equations 13-10 and 13-11; corresponding relationships are not true for an arbitrary choice of the reference point A. The generalization will be undertaken in Chapter 16.

Example

If the cylinder of the example on pages 141-142 is released from rest at the position $\phi = \phi_0$, what will be its speed as it passes a lower point on the surface?

Having found the position as a function of time, we can, of course, find the desired speed by differentiating and substituting the appropriate value of t. However, application of the energy equation will obviate the necessity of knowing $\phi(t)$ and, at the same time, give us an answer that will not depend on the assumption that ϕ is small.

Figure 13-6 shows the cylinder in an arbitrary position θ and an increment of displacement of its mass center. Since the point of contact with the surface is the instantaneous center, the increment of work done

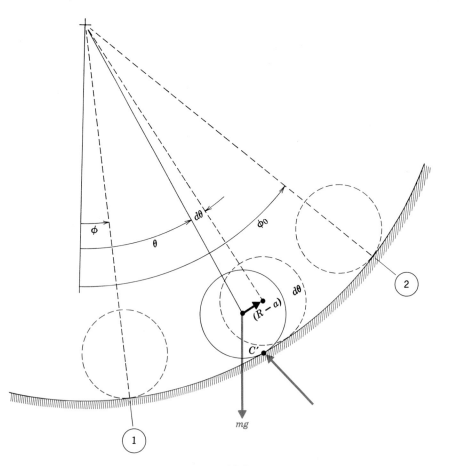

Figure 13-6

by the reaction there is zero. The increment of work done by the resultant of the gravitational forces is

$$dW = -mg(R - a) \sin \theta \, d\theta$$

The work done by these forces as the cylinder moves from position 1 to position 2 is thus

$$W = -mg(R - a) \int_{\phi}^{\phi_0} \sin \theta \, d\theta$$

$$= -mg(R - a)(\cos \phi - \cos \phi_0)$$

(Observe that since the gravitational forces are conservative, we could as well have written this last equation directly, as the difference in potential between positions 1 and 2.) Since the cylinder is at rest at $\phi = \phi_0$, the kinetic energy there is zero, so that

$$T_2 - T_1 = 0 - \frac{1}{2} I_{C'} \, \omega^2$$

The work-energy relationship then becomes

$$-mg(R - a)(\cos \phi - \cos \phi_0) = -\frac{1}{2} \left(\frac{3ma^2}{2}\right)\left[\left(\frac{R - a}{a}\right) \dot{\phi}\right]^2$$

from which the speed at any position ϕ is easily determined:

$$\dot{\phi}^2 = \frac{4g}{3(R - a)} (\cos \phi - \cos \phi_0)$$

This relationship can be used to determine the angle ϕ (not necessarily small) as a function of time, by proceeding as for the spring-mass system in Section 11-3.

Problems

13-42 (a) A torsional spring has stiffness α (N·m/rad). Compute the work done on the spring by M as it twists the shaft slowly from zero to an angle θ.

(b) A disc having moment of inertia I (kg·m²) is attached to the spring. It is twisted an amount θ_0 and released suddenly. Write the equation of energy conservation, solve for the angular velocity $\dot{\theta}$, separate the variables t and θ as were the variables t and x in pp. 66–67, and determine $\theta(t)$.

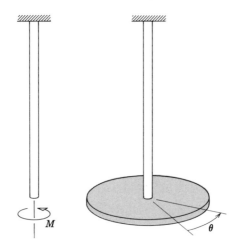

13-43   The homogeneous circular cylinder rolls off the circular cylindrical surface without slipping. If it is nudged from the equilibrium position with negligible velocity, at what point will it leave the surface?

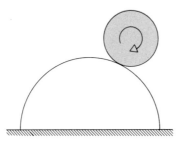

13-44   If a homogeneous sphere rolls off a circular cylinder surface in the same manner as the cylinder of the previous problem, at what point will it leave the surface?

13-45   Write the equation of conservation of mechanical energy for the system in Problem 13-21. Differentiate this equation to obtain the acceleration of the racer.

13-46   The mass of the connecting rod in Figure 10-12 is m, its radius of gyration about point P is k_p, and the distance from P to its center of mass is c. Evaluate its kinetic energy as a function of θ and $\dot{\theta}$.

13-47   The gear has a mass of 12 kg and a radius of gyration (with respect to the rotation axis through its mass center) of 130 mm. The rod has a mass of 9 kg. The spring has a stiffness of 350 N/m and an unstretched length of 460 mm. The system is released from rest with $\theta = 30°$. Estimate the maximum velocity of the center of the gear.

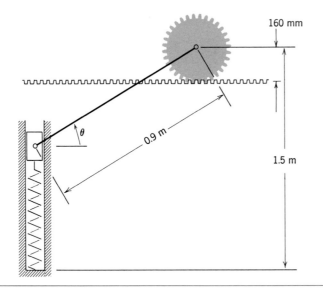

13-5

ANGULAR IMPULSE-ANGULAR MOMENTUM. The impulse-momentum law (12-5) involving the translational motion of the mass center applies here as well as for more general systems of particles.

$$\int_1^2 \mathbf{f}\, dt = m(\mathbf{v}_{C2} - \mathbf{v}_{C1})$$

Integration of each member of Equation 13-5 with respect to time, between limits of t_1 and t_2, gives us the analogous relationship governing the rotational part of the motion.

$$\int_1^2 M_{Az}\, dt = I_{A\omega}(\omega_2 - \omega_1) + m\mathbf{u}_z \cdot \int_1^2 \mathbf{r}_{C/A} \times d\mathbf{v}_A \qquad (13\text{-}12)$$

The left-hand member of this equation is the z component of a vector quantity called the *angular impulse*. The first term on the right side of the equation may be recognized as the change in the z component of angular momentum. When the reference point A is chosen so that the term $m\mathbf{r}_{C/A} \times \mathbf{a}_A$ in Equation 13-5 vanishes, we can say that *the angular impulse is equal to the change in angular momentum.*

Example

The homogeneous circular cylinder shown in Figure 13-7a rolls along the flat surface until it strikes the "curb." If the cylinder rolls over the corner

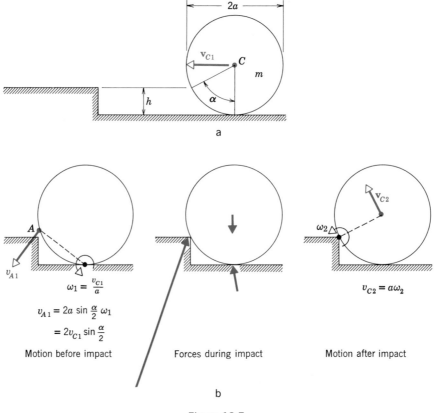

Figure 13-7

without slipping or bouncing off, at what speed will the center be traveling immediately after impact?

Observe that the center of mass will undergo a significant change in velocity during a very short period of time; the very large acceleration will thus be accompanied by a correspondingly large reactive force at the corner. While rigid body analysis will be incapable of telling us anything about how this force varies during impact, the impulse-momentum relationship, covering a short time interval including the duration of the large reaction, can give us an evaluation of the motion after impact. The pertinent aspects of the motion before and after impact are shown in Figure 13-7b.

Let us choose the reference point A at the corner, so that the large unknown reactive force does not enter Equation 13-12. Since the time interval of integration is very small and the moment of forces about point A is not correspondingly large,* the angular impulse is negligible:

$$\int_1^2 M_A \, dt \approx 0$$

The change in angular momentum is simply

$$I_A(\omega_2 - \omega_1) = \frac{3ma^2}{2} \left(\frac{v_{C2}}{a} - \frac{v_{C1}}{a} \right)$$

The remaining term in Equation 13-12 can be evaluated after noting that during the time interval of integration, the position vector $\mathbf{r}_{C/A}$ is essentially fixed. Thus,

$$m\mathbf{u}_z \cdot \int_1^2 \mathbf{r}_{C/A} \times d\mathbf{v}_A = m\mathbf{u}_z \cdot \left(\mathbf{r}_{C/A} \times \int_1^2 d\mathbf{v}_A \right)$$

$$= m\mathbf{u}_z \cdot [\mathbf{r}_{C/A} \times (\mathbf{0} - \mathbf{v}_{A1})]$$

With reference to Figure 13-7b, this last expression is evaluated as follows.

$$m\mathbf{u}_z \cdot [\mathbf{r}_{C/A} \times (-\mathbf{v}_{A1})] = mav_{A1} \sin \frac{\alpha}{2}$$

$$= 2ma \sin^2 \frac{\alpha}{2} v_{C1}$$

$$= ma(1 - \cos \alpha)v_{C1}$$

$$= mhv_{C1}$$

Substitution of the quantities evaluated above into Equation 13-12 gives us a relationship in which v_{C2} is the only unknown quantity.

$$0 = \frac{3ma}{2} (v_{C2} - v_{C1}) + mhv_{C1}$$

$$v_{C2} = \left(1 - \frac{2h}{3a} \right) v_{C1}$$

We note that the cylinder suffers a loss of speed during the impact process. Other interesting aspects of this process are brought out in several of the problems.

---

* "Correspondingly large" here means of sufficient magnitude that the integral $\int M_A \, dt$ is of the same order of magnitude as the other terms in Equation 13-12. If, for example, A were taken somewhere such that M involved the large reactive force at the corner, the angular impulse would not be negligible.

Problems

13-48 The bar rotates with the disc in the position shown, when the disc is suddenly stopped. The moment of inertia of the bar about the pivot is I_A. Determine the angular velocity of the bar after the disc is stopped. Also, evaluate the kinetic energy in the bar before and after stoppage.

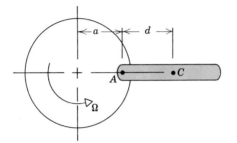

13-49 The 160-grain bullet leaves the muzzle of the 8-lb, 12 oz rifle at 3150 ft/s. Also traveling down the barrel are some 70 grains of gaseous combustion products having an average velocity of about half that of the bullet.

0.85 in

Before any appreciable recoil is absorbed by the firer, what are the velocity of the center of mass of the rifle and the angular velocity of the rifle? What are the kinetic energies of (a) the bullet, (b) translation of the rifle, and (c) rotation of the rifle?

13-50 Rework the example on pp. 162–163 using the center of mass as a reference for angular impulse and momentum.

13-51 Evaluate the impulse of the large reactive force at the curb of the example on pp. 162–163.

13-52 What fraction of the kinetic energy of the cylinder of the example on pp. 162–163 is lost during impact?

13-53 At what speed will the wheel be rolling on the upper level in the example on pp. 162–163?

APPENDIX A
SOME USEFUL
NUMERICAL VALUES

A-1

PHYSICAL CONSTANTS

Universal gravitational constant	6.67×10^{-11} N·m²/kg²
Mass of the earth	5.983×10^{-4} kg
Speed of light	0.2998 Gm/s
Standard atmospheric pressure	101.325 kPa
Density of air (0°C, 1 atm)	1.29 kg/m³
Density of water (4°C, 1 atm)	1.0000 Mg/m³
Density of concrete	2.4 Mg/m³
Density of steel	7.7–7.9 Mg/m³

A-2

PREFIXES FOR SI UNITS. Exponents in higher-order derived units apply to prefixes as well as units. For example, km² = (10³ m)² = 10⁶ m²·

Prefix	SI Symbol	Multiplication Factor
tera	T	10^{12}
giga	G	10^{9}
mega	M	10^{6}
kilo	k	10^{3}
hecto*	h*	10^{2}
deka*	da*	10^{1}
deci*	d*	10^{-1}
centi*	c*	10^{-2}
milli	m	10^{-3}
micro	μ	10^{-6}
nano	n	10^{-9}
pico	p	10^{-12}
femto	f	10^{-15}
atto	a	10^{-18}

* Use discouraged.

A-3

UNITS OF MEASUREMENT

Quantity	SI Unit	Other Units
length	meter(m)	inch = 25.400 mm
		foot = 0.304 800 m
		statute mile = 1.609 344 km
		nautical mile = 1.852 000 km
		fathom = 1.828 800 m
mass	kilogram (kg)	pound-mass = 0.453 592 37 kg
		slug = 14.593 903 kg
		grain = 64.798 910 mg
time	second (s)	minute = 60 s
		hour = 3600 s
temperature	kelvin (K)	degrees Fahrenheit $t_K = (t_F + 459.67)/1.8$
		degrees Celsius $t_K = t_C + 273.15$
plane angle	radian (rad)	degree = $(\pi/180)$ rad
		minute = $(1/160)$ deg
		second = $(1/60)$ min
angular velocity	rad/s	rpm = $(30/\pi)$ rad/s
area	m²	acre = 4046.856 m²
		hectare = 10^4 m²
energy	Joule (J)	foot-pound-force = 1.355 818 J
	$J = N \cdot m$	erg = 10^{-7} J
	$= kg \cdot m^2/s^2$	calorie (mean) = 4.190 02 J
		British thermal unit = 1.055 kJ
force	newton (N)	pound-force = 4.448 222 N
	$N = kg \cdot m/s^2$	kip = 4.448 222 kN
		poundal = 0.138 255 N
		dyne = 10^{-5} N
		kilogram-force = 9.806 650 N
power	watt (W)	horsepower (550 ft-lbf/s) = 745.7 W
	$W = J/s$	British thermal unit per hour = 0.293 071 W
	$= kg \cdot m^2/s^3$	refrigeration ton = 3.517 kW
pressure	pascal (Pa)	bar = 100 kPa
	$Pa = N/m^2$	pound per square inch = 6.894 757 kPa
	$= kg/m \cdot s^2$	centimeter of mercury (0°C) = 1.333 22 kPa
		centimeter of water (4°C) = 98.0638 Pa
velocity	m/s	kilometer per hour = 0.277 778 m/s
		mile per hour = 0.447 040 m/s
		knot = 0.514 444 m/s
volume	m³	liter = dm³ = 10^{-3} m³
		fluid ounce = $2.957\ 353 \times 10^{-5}$ m³
		U.S. liquid gallon = $3.785\ 412 \times 10^{-3}$ m³
		barrel = 0.158 987 m³

APPENDIX B
SOME FORMULAS
OF VECTOR ANALYSIS

B-1

DEFINITIONS. A vector **A** specifies magnitude and direction, but no more. In order to specify "line of action" or "point of application" of a vector (such as for a force vector), an additional vector, a position vector, is used.

Magnitude: $A = |\mathbf{A}|$

$\mathbf{A}_x$ = projection of **A** onto x.

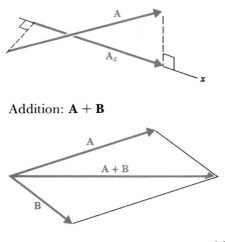

Addition: $\mathbf{A} + \mathbf{B}$

167

Negative: $-\mathbf{A}$

Subtraction: $\mathbf{A} - \mathbf{B} = \mathbf{A} + (-\mathbf{B})$

Multiplication by scalar:

$$|p\mathbf{A}| = |p||\mathbf{A}|$$

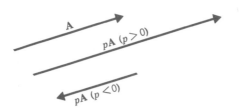

Dot product: $\mathbf{A} \cdot \mathbf{B} = AB \cos \sphericalangle{}_{\mathbf{A}}^{\mathbf{B}}$

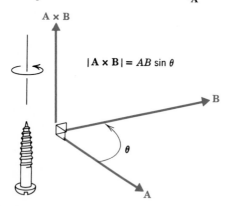

Cross product:

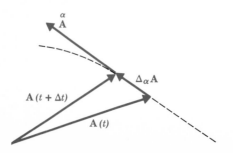

Derivative:

$$\overset{\alpha}{\dot{\mathbf{A}}} = \underset{\Delta t \to 0}{\mathrm{Lim}} \frac{\Delta_\alpha \mathbf{A}}{\Delta t}$$

where $\Delta_\alpha \mathbf{A}$ is the change, observed in the α reference frame, of the vector $\mathbf{A}$, corresponding to the change in t, Δt.

Rectangular Cartesian components:

$$\mathbf{A} = A_x \mathbf{u}_x + A_y \mathbf{u}_y + A_z \mathbf{u}_z$$

$$\mathbf{u}_i = \frac{\mathbf{A}_i}{A_i}$$

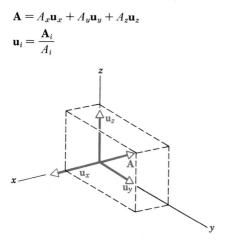

B–2

FORMULAS

$$\mathbf{A} \cdot \mathbf{B} = \mathbf{B} \cdot \mathbf{A} \qquad\qquad [3\text{-}7]$$

$$(\mathbf{A} + \mathbf{B}) \cdot \mathbf{C} = \mathbf{A} \cdot \mathbf{C} + \mathbf{B} \cdot \mathbf{C} \qquad\qquad [3\text{-}8]$$

$$\mathbf{A} \times \mathbf{B} = -\mathbf{B} \times \mathbf{A} \qquad\qquad [3\text{-}10]$$

$$\mathbf{A} \times (\mathbf{B} + \mathbf{C}) = \mathbf{A} \times \mathbf{B} + \mathbf{A} \times \mathbf{C} \qquad\qquad [3\text{-}11]$$

$$\mathbf{A} \times (\mathbf{B} \times \mathbf{C}) = (\mathbf{C} \cdot \mathbf{A})\mathbf{B} - (\mathbf{A} \cdot \mathbf{B})\mathbf{C} \qquad\qquad [3\text{-}13]$$

$$\mathbf{A} \cdot (\mathbf{B} \times \mathbf{C}) = \mathbf{B} \cdot (\mathbf{C} \times \mathbf{A}) = \mathbf{C} \cdot (\mathbf{A} \times \mathbf{B}) \qquad\qquad [3\text{-}14]$$

$$\mathbf{A} = \frac{(\mathbf{B} \cdot \mathbf{A})\mathbf{B}}{\mathbf{B} \cdot \mathbf{B}} + \frac{(\mathbf{B} \times \mathbf{A}) \times \mathbf{B}}{\mathbf{B} \cdot \mathbf{B}} \qquad\qquad [3\text{-}15]$$

$$\mathbf{A} \cdot \mathbf{B} = A_x B_x + A_y B_y + A_z B_z \qquad\qquad [3\text{-}20]$$

$$\mathbf{A} \times \mathbf{B} = (A_y B_z - A_z B_y)\mathbf{u}_x$$
$$+ (A_z B_x - A_x B_z)\mathbf{u}_y$$
$$+ (A_x B_y - A_y B_x)\mathbf{u}_z \qquad\qquad [3\text{-}22]$$

$$\mathbf{A} \cdot (\mathbf{B} \times \mathbf{C}) = \begin{vmatrix} A_x & A_y & A_z \\ B_x & B_y & B_z \\ C_x & C_y & C_z \end{vmatrix} \qquad\qquad [3\text{-}23]$$

$$\overset{\alpha}{\dot{\mathbf{A}}} = \dot{A}_x \mathbf{u}_x + \dot{A}_y \mathbf{u}_y + \dot{A}_z \mathbf{u}_z$$
provided the $\mathbf{u}_i$ are fixed in the reference frame α.

$$\overset{\alpha}{\overbrace{(p\mathbf{A})}} = \dot{p}\mathbf{A} + p\overset{\alpha}{\dot{\mathbf{A}}} \qquad\qquad [9\text{-}6]$$

$$\overset{\alpha}{\overbrace{(\mathbf{A} \cdot \mathbf{B})}} = \overset{\alpha}{\dot{\mathbf{A}}} \cdot \mathbf{B} + \mathbf{A} \cdot \overset{\alpha}{\dot{\mathbf{B}}} \qquad\qquad [9\text{-}7]$$

$$\overset{\alpha}{\overbrace{(\mathbf{A} \times \mathbf{B})}} = \overset{\alpha}{\dot{\mathbf{A}}} \times \mathbf{B} + \mathbf{A} \times \overset{\alpha}{\dot{\mathbf{B}}} \qquad\qquad [9\text{-}8]$$

APPENDIX C
PROPERTIES OF LINES, AREAS, VOLUMES, AND SOLIDS

LINES

$$L = \text{length} \qquad c_i = \frac{1}{L} \int i \, dL \qquad (i = x, y, z)$$

$$(c_x, c_y, c_z) = \text{coordinates of centroid}$$

	Segment of circle	Parabola
$L =$	$2a\alpha$	$\dfrac{a}{2}\left[u + \dfrac{a}{2b} \log\left(u + \dfrac{2b}{a}\right)\right]$ $u = \sqrt{1 + (2b/a)^2}$
$c_x =$	$\dfrac{a \sin \alpha}{\alpha}$	$\dfrac{a^4}{12b^2 L}(u^3 - 1)$ $u = \sqrt{1 + (2b/a)^2}$
$c_y =$	0	$\dfrac{a^3}{32bL}\{[1 + 2(2b/a)^2]u$ $\quad - (a/2b)\log(u + 2b/a)\}$ $u = \sqrt{1 + (2b/a)^2}$

C-2

PLANE AREAS

$$A = \text{Area} \qquad c_i = \frac{1}{A} \int i \, dA \qquad (i = x, y)$$

$$A_{xx} = \int y^2 \, dA \qquad A_{yy} = \int x^2 \, dA \qquad A_{xy} = -\int xy \, dA$$

The second moments of area, denoted here as A_{ij}, were denoted as I_{ij} in *Statics*. In this volume, I_{ij} denotes second moment of mass.

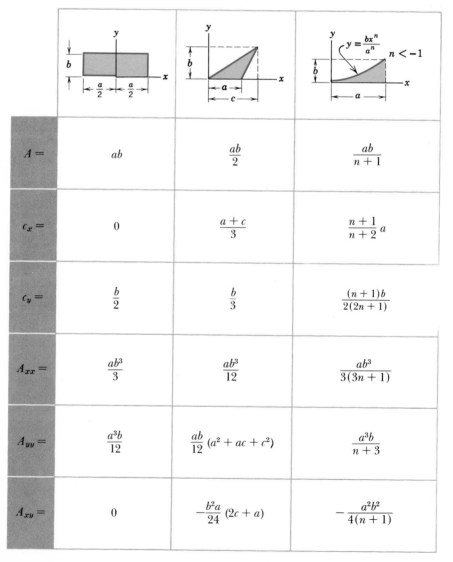

$A =$	ab	$\dfrac{ab}{2}$	$\dfrac{ab}{n+1}$
$c_x =$	0	$\dfrac{a+c}{3}$	$\dfrac{n+1}{n+2}\, a$
$c_y =$	$\dfrac{b}{2}$	$\dfrac{b}{3}$	$\dfrac{(n+1)b}{2(2n+1)}$
$A_{xx} =$	$\dfrac{ab^3}{3}$	$\dfrac{ab^3}{12}$	$\dfrac{ab^3}{3(3n+1)}$
$A_{yy} =$	$\dfrac{a^3b}{12}$	$\dfrac{ab}{12}(a^2 + ac + c^2)$	$\dfrac{a^3b}{n+3}$
$A_{xy} =$	0	$-\dfrac{b^2a}{24}(2c+a)$	$-\dfrac{a^2b^2}{4(n+1)}$

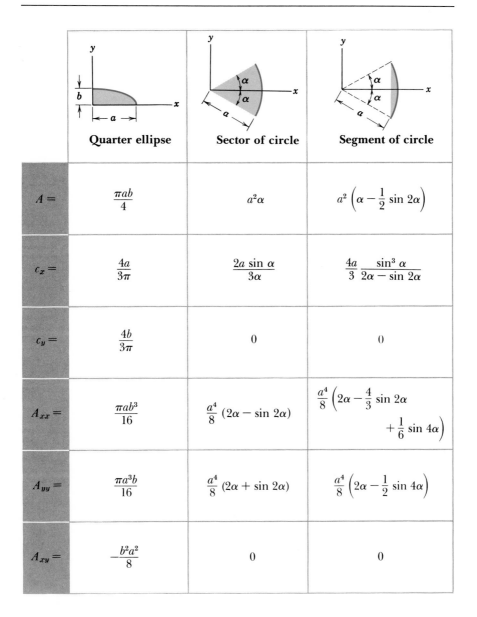

	Quarter ellipse	Sector of circle	Segment of circle
$A =$	$\dfrac{\pi ab}{4}$	$a^2\alpha$	$a^2\left(\alpha - \dfrac{1}{2}\sin 2\alpha\right)$
$c_x =$	$\dfrac{4a}{3\pi}$	$\dfrac{2a\sin\alpha}{3\alpha}$	$\dfrac{4a}{3}\dfrac{\sin^3\alpha}{2\alpha - \sin 2\alpha}$
$c_y =$	$\dfrac{4b}{3\pi}$	0	0
$A_{xx} =$	$\dfrac{\pi ab^3}{16}$	$\dfrac{a^4}{8}(2\alpha - \sin 2\alpha)$	$\dfrac{a^4}{8}\left(2\alpha - \dfrac{4}{3}\sin 2\alpha + \dfrac{1}{6}\sin 4\alpha\right)$
$A_{yy} =$	$\dfrac{\pi a^3 b}{16}$	$\dfrac{a^4}{8}(2\alpha + \sin 2\alpha)$	$\dfrac{a^4}{8}\left(2\alpha - \dfrac{1}{2}\sin 4\alpha\right)$
$A_{xy} =$	$-\dfrac{b^2 a^2}{8}$	0	0

C–3

VOLUMES

$$V = \text{Volume} \qquad c_i = \frac{1}{V} \int i \, dV \qquad (i = x, y, z)$$

$$(c_x, c_y, c_z) = \text{coordinates of centroid}$$

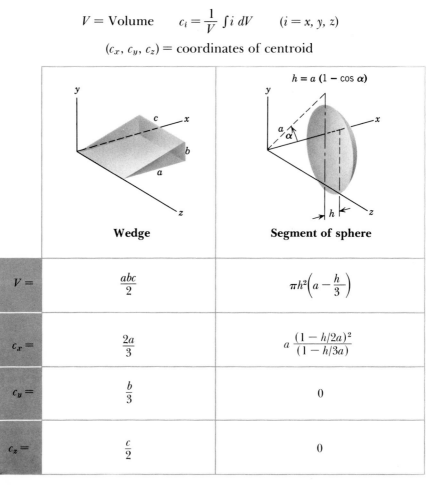

	Wedge	Segment of sphere $h = a(1 - \cos \alpha)$
$V =$	$\dfrac{abc}{2}$	$\pi h^2 \left(a - \dfrac{h}{3} \right)$
$c_x =$	$\dfrac{2a}{3}$	$a \, \dfrac{(1 - h/2a)^2}{(1 - h/3a)}$
$c_y =$	$\dfrac{b}{3}$	0
$c_z =$	$\dfrac{c}{2}$	0

$$V = \frac{A\,h}{3}$$

$$c_x = \frac{3}{4} b_x$$

$$c_y = \frac{3}{4} b_y$$

$$c_z = \frac{1}{4} h$$

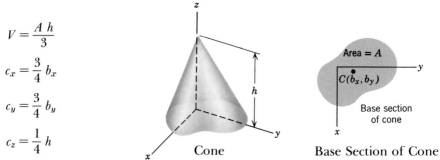

Cone

Base Section of Cone

C-4

SECOND MOMENTS OF MASS OF SOME HOMOGENEOUS-BODIES

$$I_{xx} = \int_m (y^2 + z^2)\, dm \qquad I_{yy} = \int_m (z^2 + x^2)\, dm \qquad I_{zz} = \int_m (x^2 + y^2)\, dm$$

$$I_{xy} = -\int_m xy\, dm \qquad I_{yz} = -\int_m yz\, dm \qquad I_{zx} = -\int_m zx\, dm$$

Circular Segment of Slender Rod

$$I_{xx} = \frac{ma^2}{2}\left(1 - \frac{\sin 2\alpha}{2\alpha}\right)$$

$$I_{yy} = ma^2$$

$$I_{zz} = \frac{ma^2}{2}\left(1 + \frac{\cos 2\alpha}{2\alpha}\right)$$

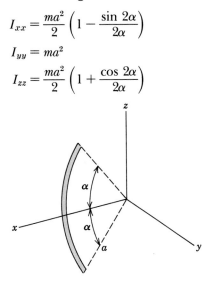

Straight Slender Rod

$$I_{XX} = I_{YY} = \frac{mL^2}{12}$$

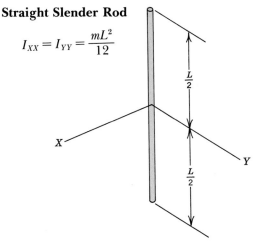

Thin Flat Plate

$$I_{ij} = \frac{m \, A_{ij}}{A} \qquad (i, j = x, y)$$

$$I_{zz} = I_{xx} + I_{yy}$$

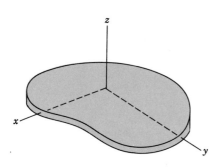

$A = $ Area of plate

$A_{ij} = $ Second moments of area

Cylinder

$$I_{xx} = m \left(\frac{A_{xx}}{A} + \frac{L^2}{12} \right)$$

$$I_{xy} = m \, \frac{A_{xy}}{A}$$

$$I_{yy} = m \left(\frac{A_{yy}}{A} + \frac{L^2}{12} \right)$$

$$I_{zz} = m \, \frac{A_{xx} + A_{yy}}{A}$$

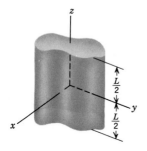

$A = $ Area of cross section

$A_{ij} = $ Second moments of area of cross section

Cone

$$I_{xx} = m \left(\frac{3A_{xx}}{5A} + \frac{h^2}{10} \right)$$

$$I_{xy} = m \frac{3A_{xy}}{5A}$$

$$I_{xz} = -\frac{3mb_x h}{20}$$

$$I_{yy} = m \left(\frac{3A_{yy}}{5A} + \frac{h^2}{10} \right)$$

$$I_{yz} = -\frac{3mb_y h}{20}$$

$$I_{zz} = \frac{3m}{5} \left(\frac{A_{xx} + A_{yy}}{A} \right)$$

A = Area of base section

(b_x, b_y) = Coordinates of centroid of base section

A_{ij} = Second moments of area of base section

Ellipsoid

$$I_{XX} = \tfrac{1}{5} m (b^2 + c^2)$$

$$I_{YY} = \tfrac{1}{5} m (c^2 + a^2)$$

$$I_{ZZ} = \tfrac{1}{5} m (a^2 + b^2)$$

Segment of Sphere

$$I_{xx} = \frac{2mah}{3}\left(\frac{1 - \dfrac{3h}{4a} + \dfrac{3h^2}{20a^2}}{1 - \dfrac{h}{3a}}\right)$$

$$I_{yy} = ma^2\left(\frac{1 - \dfrac{4h}{3a} + \dfrac{3h^2}{4a^2} - \dfrac{3h^3}{20a^3}}{1 - \dfrac{h}{3a}}\right)$$

$$h = a\,(1 - \cos \alpha)$$

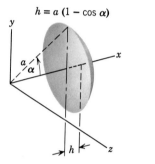

REFERENCES

Metric Practice Guide (American National Standards Institute Booklet Z 210.1, 1974).

Isaac Asimov, *Asimov's Biographical Encyclopedia of Science and Technology* (Doubleday and Company, 1972).

George W. Housner and D. E. Hudson, *Applied Mechanics Dynamics,* Second Edition (D. Van Nostrand Company, 1959).

Edward J. Routh, *Dynamics of a Particle* (G. E. Stechert and Company, 1898).

Edward J. Routh, *Dynamics of a System of Rigid Bodies, Part I,* Seventh Edition (Macmillan and Company, 1905).

Edward J. Routh, *Dynamics of a System of Rigid Bodies, Part II,* Sixth Edition (Macmillan and Company, 1905).

Murray R. Spiegel, *Vector Analysis* (Schaum's Outline Series, Schaum Publishing Company, 1959).

E. T. Whittaker, *A Treatise on the Analytical Dynamics of Particles and Rigid Bodies,* Fourth Edition (Cambridge University Press, 1961).

INDEX